P9-DFI-048

280
Topics in Current Chemistry

Topics in Current Chemistry

Recently Published and Forthcoming Volumes

Photochemistry and Photophysics of Coordination Compounds II
Volume Editors: Balzani, C., Campagna, S.
Vol. 281, 2007

Photochemistry and Photophysics of Coordination Compounds I
Volume Editors: Balzani, C., Campagna, S.
Vol. 280, 2007

Metal Catalyzed Reductive C–C Bond Formation
A Departure from Preformed Organometallic Reagents
Volume Editor: Krische, M. J.
Vol. 279, 2007

Combinatorial Chemistry on Solid Supports
Volume Editor: Bräse, S.
Vol. 278, 2007

Creative Chemical Sensor Systems
Volume Editor: Schrader, T.
Vol. 277, 2007

In situ NMR Methods in Catalysis
Volume Editors: Bargon, J., Kuhn, L. T.
Vol. 276, 2007

Sulfur-Mediated Rearrangements II
Volume Editor: Schaumann, E.
Vol. 275, 2007

Sulfur-Mediated Rearrangements I
Volume Editor: Schaumann, E.
Vol. 274, 2007

Bioactive Conformation II
Volume Editor: Peters, T.
Vol. 273, 2007

Bioactive Conformation I
Volume Editor: Peters, T.
Vol. 272, 2007

Biomineralization II
Mineralization Using Synthetic Polymers and Templates
Volume Editor: Naka, K.
Vol. 271, 2007

Biomineralization I
Crystallization and Self-Organization Process
Volume Editor: Naka, K.
Vol. 270, 2007

Novel Optical Resolution Technologies
Volume Editors:
Sakai, K., Hirayama, N., Tamura, R.
Vol. 269, 2007

Atomistic Approaches in Modern Biology
From Quantum Chemistry to Molecular Simulations
Volume Editor: Reiher, M.
Vol. 268, 2006

Glycopeptides and Glycoproteins
Synthesis, Structure, and Application
Volume Editor: Wittmann, V.
Vol. 267, 2006

Microwave Methods in Organic Synthesis
Volume Editors: Larhed, M., Olofsson, K.
Vol. 266, 2006

Supramolecular Chirality
Volume Editors: Crego-Calama, M., Reinhoudt, D. N.
Vol. 265, 2006

Photochemistry and Photophysics of Coordination Compounds I

Volume Editors: Vincenzo Balzani · Sebastiano Campagna

With contributions by

G. Accorsi · N. Armaroli · V. Balzani · G. Bergamini · S. Campagna
F. Cardinali · C. Chiorboli · M. T. Indelli · N. A. P. Kane-Maguire
A. Listorti · F. Nastasi · F. Puntoriero · F. Scandola

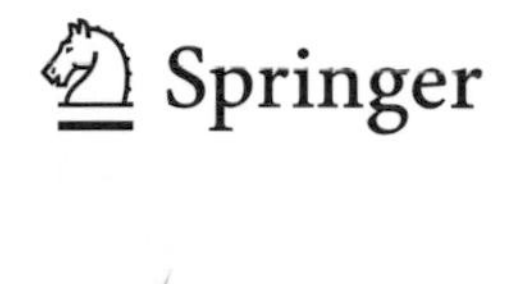

The series *Topics in Current Chemistry* presents critical reviews of the present and future trends in modern chemical research. The scope of coverage includes all areas of chemical science including the interfaces with related disciplines such as biology, medicine and materials science. The goal of each thematic volume is to give the nonspecialist reader, whether at the university or in industry, a comprehensive overview of an area where new insights are emerging that are of interest to a larger scientific audience.
As a rule, contributions are specially commissioned. The editors and publishers will, however, always be pleased to receive suggestions and supplementary information. Papers are accepted for *Topics in Current Chemistry* in English.
In references *Topics in Current Chemistry* is abbreviated Top Curr Chem and is cited as a journal.
Visit the TCC content at springerlink.com

ISSN 0340-1022
ISBN 978-3-540-73346-1 Springer Berlin Heidelberg New York
DOI 10.1007/978-3-540-73347-8

Springer is a part of Springer Science+Business Media

springer.com

Cover design: WMXDesign GmbH, Heidelberg
Typesetting and Production: LE-TEX Jelonek, Schmidt & Vöckler GbR, Leipzig

Printed on acid-free paper 02/3180 YL – 5 4 3 2 1 0

Volume Editors

Prof. Vincenzo Balzani

Dipartimento di Chimica "G. Ciamician"
Università di Bologna
Via Selmi 2
40126 Bologna
Italy
vincenzo.balzani@unibo.it

Prof. Sebastiano Campagna

Dipartimento di Chimica Inorganica
Chimica Analitica e Chimica Fisica
University of Messina
Via Sperone 31
98166 Vill. S.Agata, Messina
Italy
campagna@unime.it

Editorial Board

Topics in Current Chemistry Also Available Electronically

For all customers who have a standing order to Topics in Current Chemistry, we offer the electronic version via SpringerLink free of charge. Please contact your librarian who can receive a password or free access to the full articles by registering at:

springerlink.com

If you do not have a subscription, you can still view the tables of contents of the volumes and the abstract of each article by going to the SpringerLink Homepage, clicking on "Browse by Online Libraries", then "Chemical Sciences", and finally choose Topics in Current Chemistry.

You will find information about the

- Editorial Board
- Aims and Scope
- Instructions for Authors
- Sample Contribution

at springer.com using the search function.

Preface

Photochemistry (a term that broadly speaking includes photophysics) is a branch of modern science that deals with the interaction of light with matter and lies at the crossroads of chemistry, physics, and biology. However, before being a branch of modern science, photochemistry was (and still is today), an extremely important natural phenomenon. When God said: "Let there be light", photochemistry began to operate, helping God to create the world as we now know it. It is likely that photochemistry was the spark for the origin of life on Earth and played a fundamental role in the evolution of life. Through the photosynthetic process that takes place in green plants, photochemistry is responsible for the maintenance of all living organisms. In the geological past photochemistry caused the accumulation of the deposits of coal, oil, and natural gas that we now use as fuels. Photochemistry is involved in the control of ozone in the stratosphere and in a great number of environmental processes that occur in the atmosphere, in the sea, and on the soil. Photochemistry is the essence of the process of vision and causes a variety of behavioral responses in living organisms.

Photochemistry as a science is quite young; we only need to go back less than one century to find its early pioneer [1]. The concept of coordination compounds is also relatively young; it was established in 1892, when Alfred Werner conceived his theory of metal complexes [2]. Since then, the terms coordination compound and metal complex have been used as synonyms, even if in the last 30 years, coordination chemistry has extended its scope to the binding of all kinds of substrates [3, 4].

The photosensitivity of metal complexes has been recognized for a long time, but the photochemistry and photophysics of coordination compounds as a science only emerged in the second half of the last century. The first attempt to systematize the photochemical reactions of coordination compounds was carried out in an exhaustive monograph published in 1970 [5], followed by an authoritative multi-authored volume in 1975 [6]. These two books gained the attention of the scientific community and certainly helped several inorganic and physical chemists to enter the field and to enrich and diversify their research activities. Interestingly, 1974 marked the beginning of the series of International Symposia on the Photochemistry and Photophysics of Coordi-

Contents

Contents of Volume 281

Photochemistry and Photophysics of Coordination Compounds II

Volume Editors: Balzani, S., Campagna, V.
ISBN: 978-3-540-73348-5

Top Curr Chem (2007) 280: 1–36
DOI 10.1007/128_2007_132

Published online: 23 June 2007

Photochemistry and Photophysics of Coordination Compounds: Overview and General Concepts

Vincenzo Balzani[1] (✉) · Giacomo Bergamini[1] · Sebastiano Campagna[2] · Fausto Puntoriero[1]

[1]Dipartimento di Chimica "G. Ciamician", Università di Bologna, 40100 Bologna, Italy
vincenzo.balzani@unibo.it

[2]Dipartimento di Chimica Inorganica, Chimica Analitica, e Chimica Fisica, Università di Messina, 98166 Messina, Italy

And God said: "Let there be light";
And there was light.
And God saw that the light was good.
(*Genesis, 1, 3–4*)

Abstract Investigations in the field of the photochemistry and photophysics of coordination compounds have proceeded along several steps of increasing complexity in the last 50 years. Early studies on ligand photosubstitution and photoredox decomposition reactions of metal complexes of simple inorganic ligands (e.g., NH_3, CN^-) were followed by accurate investigations on the photophysical behavior (luminescence quantum yields and lifetimes) and use of metal complexes in bimolecular processes (energy and electron transfer). The most significant differences between Jablonski diagrams for organic molecules and coordination compounds are illustrated. A large number of complexes stable toward photodecomposition, but capable of undergoing excited-state redox processes, have been used for interconverting light and chemical energy. The rate constants of a great number of photoinduced energy- and electron-transfer processes involving coordination compounds have been measured in order to prove the validity and/or extend the scope of modern kinetic theories. More recently, the combination of supramolecular chemistry and photochemistry has led to the design and construction of supramolecular systems capable of performing light- induced functions. In this field, luminescent and/or photoredox reactive metal complexes are presently used as essential components for a bottom-up approach to the construction of molecular devices and machines. A few examples of molecular devices for processing light signals and of molecular machines powered by light energy, based on coordination compounds, are briefly illustrated.

Keywords Coordination compounds · Electron transfer · Energy transfer · Excited-state properties · Photochemistry · Supramolecular photochemistry

1 Early History

The photosensitivity of metal complexes has been known for a long time. The first paper exhibiting some scientific character was that of Scheele (1772) on the effect of light on AgCl, and photography was becoming established in several countries in the 1830s [1]. The light sensitivity of other metal complexes (particularly $Na_4[Fe(CN)_6]$) was also observed very early [2]. At the beginning of the last century the importance of photochemistry became more widely recognized, mainly due to the work and the ideas of Giacomo Ciamician [3], Professor of Chemistry at the University of Bologna. In the same period (1912–1913), modern physics introduced the concept that light absorption corresponds to the capture of a photon by a molecule. This concept, and the distinction (sometimes difficult) between primary and secondary photoprocesses, led to the definition of quantum yield. In the following years, investigations on Fe^{3+} and UO_2^{2+} complexes were performed in looking for useful chemical actinometers (see, e.g., [4]). Several quantitative works also appeared on the photochemical behavior of $[Fe(CN)_6]^{4-}$ and Co(III)–amine complexes in aqueous solution [2]. The lack of a theory on the absorption spectra and on the nature of the excited states, however, prevented any mechanistic interpretation of the observed photoreactions as well as of the few scattered reports on luminescent complexes.

After the Second World War, the interpretation of the absorption spectra started thanks to the development of the ligand field theory [5, 6] and the first attempts to rationalize the charge-transfer bands [7, 8]. Following these developments, the photochemistry of coordination compounds could take its first steps as a modern science and in a time span of 2 years four important laboratories published their first photochemical paper [9–12]. Much of the attention was focused on Cr(III) complexes, whose luminescence was also investigated in some detail [13]. Later, Co(III) complexes attracted a great deal of attention since their photochemical behavior was found to change drastically with excitation wavelength [14, 15]. A few, isolated flash photolysis investigations began to appear, but this technique remained unavailable to most inorganic photochemists for several years.

Since the late 1960s, the great development of photochemical and luminescence investigations on organic compounds led to the publication of books [16–19] illustrating fundamental photochemical concepts that were also quickly exploited for coordination compounds [2]. From that period, it became common to discuss the photochemical and photophysical behavior of a species (be it an organic molecule or a charged metal complex) on the basis of electronic configurations, selection rules, and energy level diagram, as we do today.

2 Molecular Photochemistry

Molecules are multielectron systems. Approximate electronic wavefunctions of a molecule can be written as products of one-electron wavefunctions, each consisting of an orbital and a spin part:

$$\Psi = \Phi S = \Pi_i \varphi_i s_i \,. \qquad (1)$$

The φ_is are appropriate molecular orbitals (MOs) and s_i is one of the two possible spin eigenfunctions, α or β. The orbital part of this multielectron wavefunction defines the *electronic configuration.*

We illustrate now the procedure to construct energy level diagrams, using as examples an organic molecule and a few coordination compounds.

2.1 Organic Molecules

The MO diagram for formaldehyde is shown in Fig. 1 [20]. It consists of three low-lying σ-bonding orbitals, a π-bonding orbital of the CO group, a non-bonding orbital n of the oxygen atom (highest occupied molecular orbital, HOMO), a π-antibonding orbital of the CO group (lowest unoccupied molecular orbital, LUMO), and three high-energy σ-antibonding orbitals. The

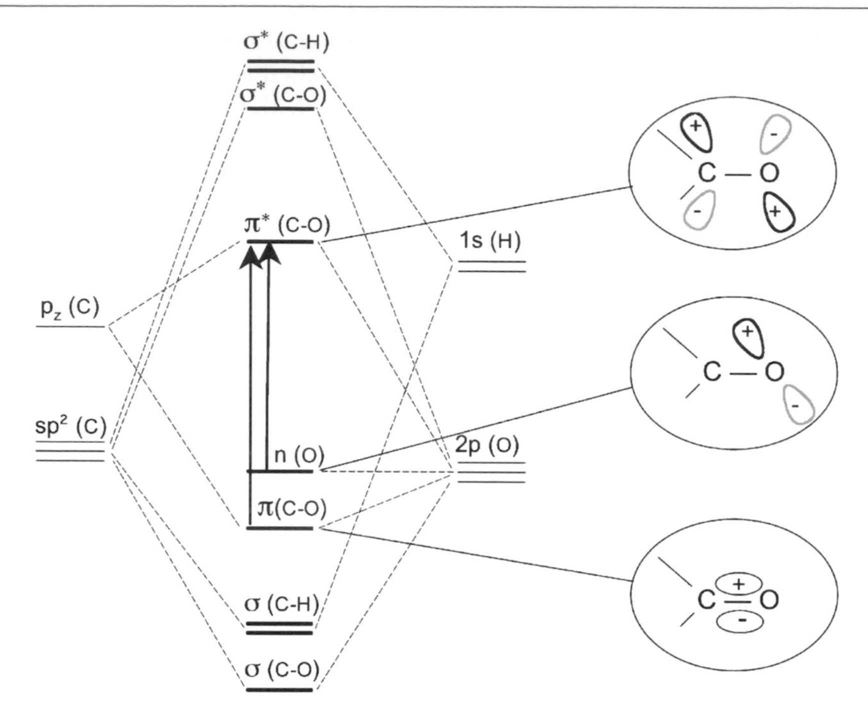

Fig. 1 Molecular orbital diagram for formaldehyde. The *arrows* indicate the $n \rightarrow \pi^*$ and $\pi \rightarrow \pi^*$ transitions

lowest-energy electronic configuration is (neglecting the filled low-energy orbitals) $\pi^2 n^2$. Excited configurations can be obtained from the ground configuration by promoting one electron from occupied to vacant MOs. At relatively low energies, one expects to find $n \rightarrow \pi^*$ and $\pi \rightarrow \pi^*$ electronic transitions (Fig. 1), leading to $\pi^2 n\pi^*$ and $\pi n^2 \pi^*$ excited configurations (Fig. 2a).

In a very crude zero-order description, the energy associated with a particular electronic configuration would be given by the sum of the energies of the occupied MOs. In order to obtain a more realistic description of the energy states of the molecule, two features should be added to the simple configuration picture: (1) spin functions must be attached to the orbital functions describing the electronic configurations, and (2) interelectronic repulsion must be taken into account. These two closely interlocked points have important consequences, since they may lead to the splitting of an electronic configuration into several states.

In the case of formaldehyde, the inclusion of spin and electronic repulsion leads to the schematic energy level diagram shown in Fig. 2b: each excited electronic configuration is split into a pair of triplet and singlet states, with the latter at higher energy because electronic repulsion is higher for spin-paired electrons. It can be noticed that the singlet–triplet splitting for the states arising from the $\pi\pi^*$ configuration is larger than that of the states cor-

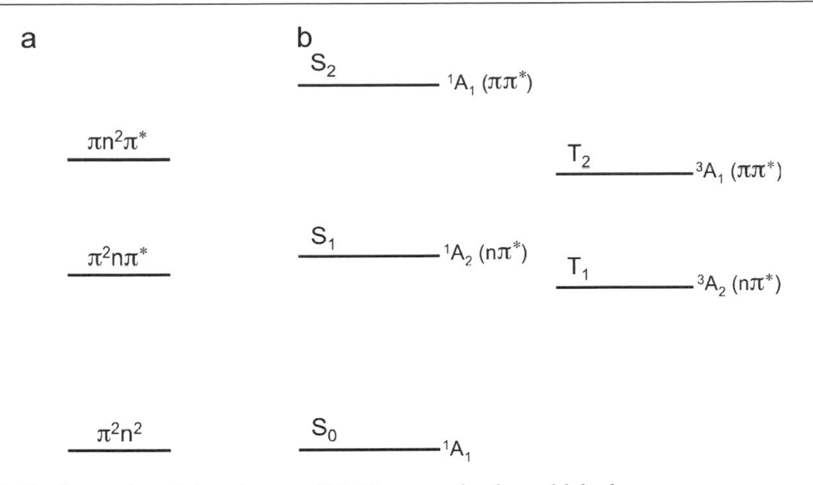

Fig. 2 Configurations (**a**) and states (**b**) diagrams for formaldehyde

responding to the $n\pi^*$ configuration. This result arises from the dependence of the interelectronic repulsions on the amount of spatial overlap between the MOs containing the two electrons, and this overlap is greater in the first than in the second case (see the MO shapes in Fig. 1). The electronic states can be designated by symbols that specify the symmetry of the wavefunction in the symmetry group of the molecule (e.g., A_1, A_2, etc. in the C_{2v} group of formaldehyde) and the spin multiplicity (number of unpaired electrons + 1) as a left superscript. In organic photochemistry, it is customary to label the singlet and triplet states as S_n and T_n, respectively, with $n = 0$ for the singlet ground state and $n = 1, 2$, etc. for states arising from the various excited configurations (often indicated in parentheses). Both notations are shown for formaldehyde in Fig. 2b. The situation sketched above (i.e., singlet ground state, pairs of singlet and triplet excited states arising from each excited configuration, lowest excited state of multiplicity higher than the ground state) is quite general for organic molecules that usually exhibit a closed-shell ground-state configuration.

State energy diagrams of this type, usually called "Jablonski diagrams", are used for the description of light absorption and of the photophysical processes that follow light excitation (vide infra).

2.2 Metal Complexes

For metal complexes, the construction of Jablonski diagrams via electronic configurations from the MO description follows the same general lines described above for organic molecules [2]. A schematic MO diagram for an octahedral transition metal complex is shown in Fig. 3. The various MOs can be conveniently classified according to their predominant atomic orbital

(Fig. 5). For most Cr(III) complexes, e.g., for $[Cr(NH_3)_6]^{3+}$, the lowest-energy transition is metal centered and the resulting $\pi_M(t_{2g})^2\sigma^*_M(e_g)$ configuration gives rise to $^4T_{2g}$ and $^4T_{1g}$ excited states (Fig. 5). Several other coordination compounds, including the complexes of the lanthanide ions, have an open-shell ground-state configuration and, as a consequence, a ground state with high-multiplicity and low-energy intraconfigurational metal-centered excited states.

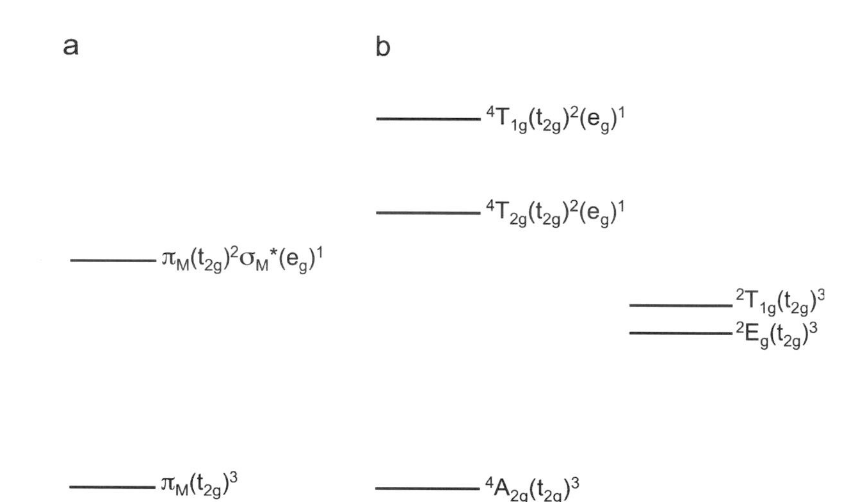

Fig. 5 Configurations (**a**) and state (**b**) diagrams for an octahedral Cr(III) complex. Only the lower-lying excited states of each configuration are shown [20]

In conclusion, metal complexes tend to have more complex and specific Jablonski diagrams than organic molecules. Points to be noticed are: (1) spin multiplicity other than singlet and triplet can occur, but for each electronic configuration the state with highest multiplicity remains the lowest one; (2) excited states can exist that belong to the same configuration of the ground state (this implies that the ground state has the highest multiplicity); and (3) more than one pair of states of different multiplicity can arise from a single electron configuration. In the following, in order to discuss some general concept of molecular photochemistry we will make use of a generic Jablonski diagram based on singlet and triplet states.

2.3 Light Absorption and Intramolecular Excited-State Decay

Figure 6 shows a schematic energy level diagram for a generic molecule [23]. In principle, transitions between states having the same multiplicity are allowed, whereas those between states of different multiplicity are forbidden. Therefore, the electronic absorption bands observed in the UV–visible spec-

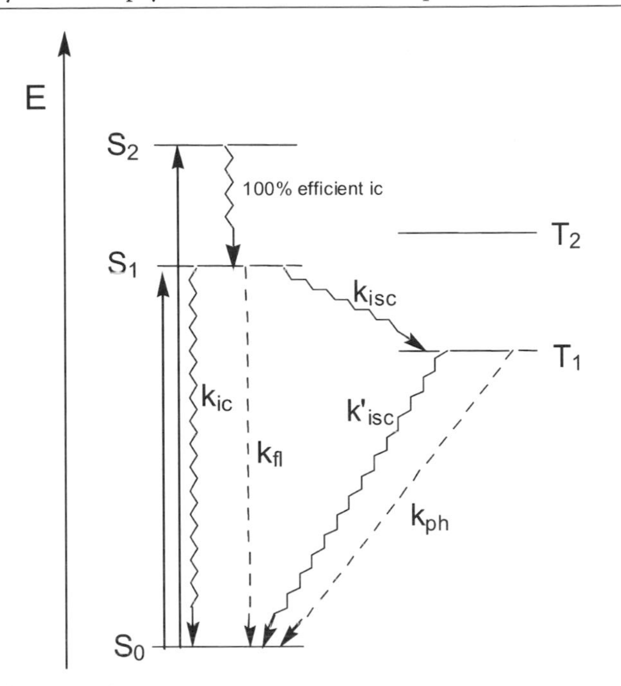

Fig. 6 Schematic energy level diagram for a generic molecule

trum of such a generic molecule would display bands corresponding to the $S_0 \rightarrow S_n$ transitions of the diagram. For metal complexes, which usually are highly symmetric species, symmetry selection rules can also play a role in determining the intensity of the absorption bands. Furthermore, the presence of a heavy atom (namely, the metal) relaxes the spin-conservation rule.

The excited states are unstable species that decay not only by intramolecular chemical reactions (e.g., dissociation, isomerization) but also (actually, more often) by intramolecular radiative and nonradiative deactivations. When a species is excited to upper spin-allowed excited states, it usually undergoes a fast and 100% efficient radiationless deactivation (internal conversion, ic) to the lowest spin-allowed excited (S_1 in Fig. 6). Setting aside the intramolecular photochemical processes, such an excited state undergoes deactivation via three competing first-order processes: nonradiative decay to the ground state (internal conversion, rate constant k_{ic}); radiative decay to the ground state (fluorescence, k_{fl}); and intersystem crossing (isc) to the lowest triplet state T_1 (k_{isc}). In its turn, T_1 can undergo deactivation via nonradiative (intersystem crossing, k'_{isc}) or radiative (phosphorescence, k_{ph}) decay to the ground state S_0. When the species contains heavy atoms, as in the case of metal complexes, the formally forbidden intersystem crossing and phosphorescence processes become faster. The lifetime (τ) of an excited state, i.e., the time needed to reduce the excited-state concentration by 2.718, is given

induced [30–34] and light-generating [35, 36] electron-transfer processes. Such studies were further boosted by the fact that, after the energy crisis of the early 1970s, several photochemists became involved in the problem of solar energy conversion. Particular interest arose around photosensitized water splitting [37–41] and it was soon realized [42] that $[Ru(bpy)_3]^{2+}$ and related complexes, because of their excited-state redox properties, might function as photocatalysts for such a process.

As a matter of fact, in the period 1975–1985 a real revolution occurred in the field of the photochemistry of coordination compounds. The study of intramolecular ligand photosubstitution, photoredox decomposition, and photoisomerization reactions was almost completely set apart, about 300 Ru(II) bipyridine-type complexes were synthesized and investigated in an attempt (mostly vain) to improve the already outstanding excited-state properties of $[Ru(bpy)_3]^{2+}$ [43], and, thanks to an extensive use of pulsed techniques, huge amounts of data were collected on the rate constants of bimolecular processes [44]. The high exergonicity of the excited-state electron-transfer reactions (and/or of their back reactions) offered the opportunity for the first time to investigate some fundamental aspects of electron-transfer theories [45], with particular attention to the so-called Marcus inverted region.

4 Supramolecular Photochemistry

4.1 Operational Definition of Supramolecular Species

In the late 1980s, following the award of the 1987 Nobel prize to Pedersen, Cram, and Lehn, there was a sudden increase of interest in supramolecular chemistry, a highly interdisciplinary field based on concepts such as molecular recognition, preorganization, and self-assembling.

The classical definition of supramolecular chemistry is that given by J.-M. Lehn, namely "the chemistry beyond the molecule, bearing on organized entities of higher complexity that result from the association of two or more chemical species held together by intermolecular forces" [46]. There is, however, a problem with this definition. With supramolecular chemistry there has been a change in focus from molecules to molecular assemblies or multicomponent systems. According to the original definition, however, when the components of a chemical system are linked by covalent bonds, the system should not be considered a supramolecular species, but a molecule. This point is particularly important in dealing with light-powered molecular devices and machines (vide infra), which are usually multicomponent systems in which the components can be linked by chemical bonds of various natures.

To better understand this point, consider, for example, the three systems [47] shown in Fig. 7, which play the role of photoinduced charge-separation molecular devices [48]. In each one of them, two components, a Zn(II) porphyrin and an Fe(III) porphyrin, can be immediately singled out. In **1**, these two components are linked by a hydrogen-bonded bridge, i.e., by intermolecular forces, whereas in **2** and **3** they are linked by covalent bonds. According to the above-reported classical definition of supramolecular chemistry, **1** is a supramolecular species, whereas **2** and **3** are (large) molecules. In each one of the three systems, the two components substantially maintain their intrinsic properties and, upon light excitation, electron transfer takes place from the Zn(II) porphyrin unit to the Fe(III) porphyrin one. The values of the rate constants for photoinduced electron transfer ($k_{el} = 8.1 \times 10^9$, 8.8×10^9, and 4.3×10^9 s^{-1} for **1**, **2**, and **3**, respectively) show that the electronic interaction between the two components in **1** is comparable to that in **2**, and is slightly stronger than that in **3**. Clearly, as far as photoinduced electron transfer is concerned, it would sound strange to say that **1** is a supramolecular species, and **2** and **3** are molecules.

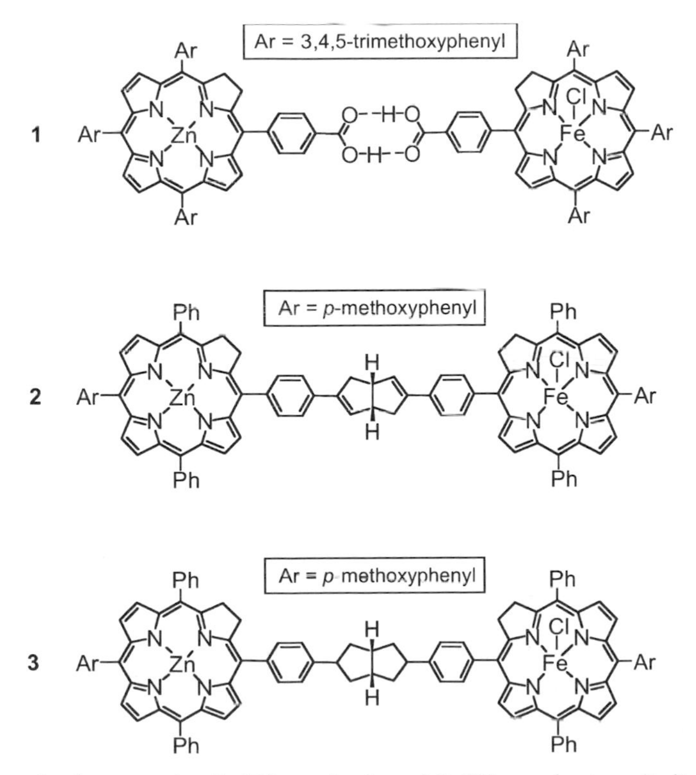

Fig. 7 Three dyads possessing Zn(II) porphyrin and Fe(III) porphyrin units linked by an H-bonded bridge (**1**), a partially unsaturated bridge (**2**), and a saturated bridge (**3**) [47]

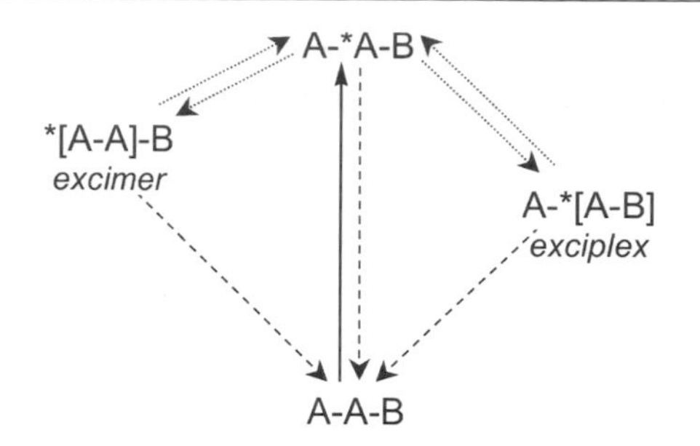

Fig. 9 Schematic representation of excimer and exciplex formation in a supramolecular system

Compared with the "monomer" emission, the emission of an excimer or exciplex is always displaced to lower energy (longer wavelengths) and usually corresponds to a broad and rather weak band.

Excimers are usually obtained when an excited state of an aromatic molecule interacts with the ground state of a molecule of the same type. For example, between the excited and ground states of anthracene units. Exciplexes are obtained when an electron donor (acceptor) excited state interacts with an electron acceptor (donor) ground-state molecule; for example, between excited states of aromatic molecules (electron acceptors) and amines (electron donors). Excited states of coordination compounds are seldom involved in excimers or exciplexes, since their components (metal and ligands) have already used their electron donor or acceptor properties in forming the complex. Furthermore, the three-dimensional structure of coordination compounds usually prevents strong electronic interaction with other species. However, for some square planar complexes excimer emission has long been reported [54] and can indeed be found for some families of Au and Pt complexes, as discussed in other chapters of this volume.

The working mechanisms of a number of biological and artificial molecular devices and machines are based on photoinduced electron- and energy-transfer processes [20, 48, 55]. Since these processes have to compete with the intrinsic decays of the relevant excited states, a key problem is that of maximizing their rates. It is therefore appropriate to summarize some basic principles of electron- and energy-transfer kinetics. [56].

4.3 Electron Transfer

4.3.1 Marcus Theory

Electron-transfer processes involving excited-state and/or ground-state molecules can be dealt with in the frame of the Marcus theory [57] and of the successive, more sophisticated theoretical models [58, 59]. Of course, when excited states are involved, the redox potential of the excited-state couple has to be used (Eqs. 9 and 10).

According to the Marcus theory [57], the rate constant for an electron-transfer process can be expressed as

$$\kappa_{el} = \nu_N \kappa_{el} \exp\left(-\frac{\Delta G^{\ddagger}}{RT}\right) , \tag{18}$$

where ν_N is the average nuclear frequency factor, κ_{el} is the electronic transmission coefficient, and $\Delta G^{\ddagger}$ is the free energy of activation. This last term can be expressed by the Marcus quadratic relationship

$$\Delta G^{\ddagger} = \frac{\lambda}{4}\left(1 + \frac{\Delta G^0}{\lambda}\right)^2 , \tag{19}$$

where ΔG^0 is the standard free energy change of the reaction and λ is the nuclear reorganizational energy (Fig. 10). This equation predicts that for a homogeneous series of reactions (i.e., for reactions having the same λ and κ_{el} values), a $\ln k_{el}$ vs ΔG^0 plot is a bell-shaped curve (Fig. 11) involving (1) a "normal" region for endoergonic and slightly exoergonic reactions, in which $\ln k_{el}$ increases with increasing driving force; (2) an activationless maximum for $\lambda \approx - \Delta G^0$; and (3) an "inverted" region for strongly exoergonic reactions, in which $\ln k_{el}$ *decreases* with increasing driving force.

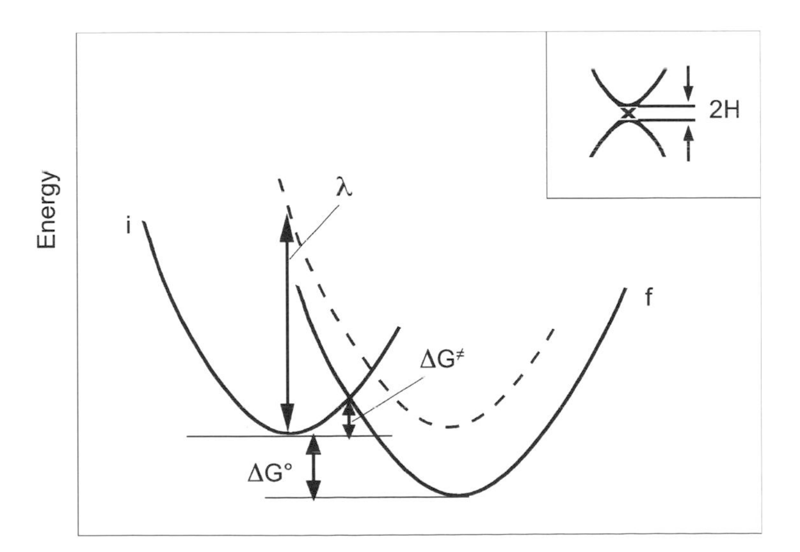

Fig. 10 Profile of the potential energy curves of an electron-transfer reaction: *i* and *f* indicate the initial and final states of the system. The *dashed curve* indicates the final state for a self-exchange (isoergonic) process. For more details, see text

where H^{el} and FC^{el} are the electronic coupling and the Franck–Condon density of states, respectively.

In the absence of any intervening medium (through-space mechanism), the electronic factor decreases exponentially with increasing distance:

$$H^{el} = H^{el}(0) \exp\left[-\frac{\beta^{el}}{2} \left(r_{AB} - r_0 \right) \right] , \tag{26}$$

where r_{AB} is the donor–acceptor distance, $H^{el}(0)$ is the interaction at the "contact" distance r_0, and β^{el} is an appropriate attenuation parameter.

For donor–acceptor components separated by vacuum, β^{el} is estimated to be in the range 2–5 Å^{-1}. When donor and acceptor are separated by "matter" (e.g., a bridge L), the electronic coupling can be mediated by mixing of the initial and final states of the system with virtual, high-energy electron-transfer states involving the intervening medium (superexchange mechanism) [60, 61].

The FC^{el} term of Eq. 25 is a thermally averaged Franck–Condon factor connecting the initial and final states. In the high temperature limit ($h\nu < k_B T$), an approximation sufficiently accurate for many room-temperature processes, the nuclear factor takes the simple form:

$$FC^{el} = \left(\frac{1}{4\pi \lambda k_B T} \right)^{1/2} \exp\left[-\frac{\left(\Delta G^0 + \lambda \right)^2}{4\lambda k_B T} \right] , \tag{27}$$

where λ is the sum of the inner (λ_i) and outer (λ_o) reorganizational energies. The exponential term of Eq. 27 is the same as that predicted by the classical Marcus model based on parabolic energy curves for initial and final states. The quantum mechanical model, however, predicts a linear, rather than a parabolic, decrease of $\ln k_{el}$ with increasing driving force in the inverted region (Fig. 11).

4.3.3 Optical Electron Transfer

Reactants and products of an electron-transfer process are intertwined by a ground/excited-state relationship. As shown in Fig. 12, for nuclear coordinates that correspond to the equilibrium geometry of A–L–B, A^+–L–B^- is an electronically excited state. Therefore, optical transitions connecting the two states are possible, as indicated by arrow 4 in Fig. 12.

The Hush theory [62] correlates the parameters that are involved in the corresponding thermal electron-transfer process by means of Eqs. 28–30:

$$E_{op} = \lambda + \Delta G^0 \tag{28}$$

$$\Delta \overline{\nu}_{1/2} = 48.06 \left(E_{op} - \Delta G^0\right)^{1/2} \tag{29}$$

$$\varepsilon_{max} \Delta \overline{\nu}_{1/2} = \left(H^{el}\right)^2 \frac{r^2}{4.20 \times 10^{-4} E_{op}} , \tag{30}$$

where E_{op}, $\Delta \overline{\nu}_{1/2}$ (both in cm^{-1}), and ε_{max} are the energy, halfwidth, and maximum intensity of the electron-transfer band, respectively, and r is the center-to-center distance. As shown by Eqs. 28–30, the energy depends on both reorganizational energy and thermodynamics, the halfwidth reflects the reorganizational energy, and the intensity of the transition is mainly related to the magnitude of the electronic coupling between the two redox centers.

In principle, therefore, important kinetic information on a thermal electron-transfer process could be obtained from the study of the corresponding optical transition. In practice, it can be shown that weakly coupled systems may undergo relatively fast electron-transfer processes without exhibiting appreciably intense optical electron-transfer bands. More details on optical electron transfer and related topics (i.e., mixed valence metal complexes) can be found in the literature [63–65].

4.4 Energy Transfer

The thermodynamic ability of an excited state to intervene in energy-transfer processes is related to its zero–zero spectroscopic energy, E^{0-0}. Bimolecular energy-transfer processes involving encounters can formally be treated using a Marcus-type approach with $\Delta G^0 = E_A^{0-0} - E_B^{0-0}$ and $\lambda \sim \lambda_i$ [66].

Energy transfer in a supramolecular system can be viewed as a radiationless transition between two "localized" electronically excited states. Therefore, the rate constant can again be obtained by an appropriate "golden rule" expression, similar to that seen above for electron transfer:

$$k_{en} = \frac{4\pi^2}{h} \left(H^{en}\right)^2 FC^{en} , \tag{31}$$

where H^{en} is the electronic coupling between the two excited states interconverted by the energy-transfer process and FC^{en} is an appropriate Franck–Condon factor. As for electron transfer, the Franck–Condon factor can be cast either in classical [67] or quantum mechanical [68–70] terms. Classically, it accounts for the combined effects of energy gradient and nuclear reorganization on the rate constant. In quantum mechanics terms, the FC factor is a thermally averaged sum of vibrational overlap integrals. Experimental information on this term can be obtained from the overlap integral

transfer should be approximately equal to the sum of the attenuation factors for two separated electron-transfer processes, i.e., β^{el} for electron transfer between the LUMOs of the donor and acceptor, and β^{ht} for the electron transfer between the HOMOs (superscript ht is for hole transfer from the donor to the acceptor). This prediction has been confirmed by experiments [74].

The spin selection rules for this type of mechanism arise from the need to obey spin conservation in the reacting pair as a whole. This allows the exchange mechanism to be operative in many cases in which the excited states involved are spin-forbidden in the usual spectroscopic sense. Thus, the typical example of an efficient exchange mechanism is that of triplet–triplet energy transfer:

$$^{*}A(T_1) - L - B(S_0) \rightarrow A(S_0) - L - {}^{*}B(T_1) \,. \qquad (38)$$

Exchange energy transfer from the lowest spin-forbidden excited state is expected to be the rule for metal complexes [61, 75].

Although the exchange mechanism was originally formulated in terms of direct overlap between donor and acceptor orbitals, it is clear that it can be extended to cover the case in which coupling is mediated by the intervening medium (i.e., the connecting bridge), as discussed above for electron-transfer processes (superexchange mechanism) [61].

5 Coordination Compounds as Components of Photochemical Molecular Devices and Machines

In the last few years, a combination of supramolecular chemistry and photochemistry has led to the design and construction of supramolecular systems capable of performing interesting light-induced functions. Photoinduced energy and electron transfer are indeed basic processes for connecting light energy inputs with a variety of optical, electrical, and mechanical functions, i.e., to obtain molecular-level devices and machines [48, 55]. We will now describe a few classical examples of molecular devices and machines in which coordination compounds are used to process light signals or to exploit light energy. Other examples are, of course, described in the chapters dealing with the complexes of the various metals.

5.1 A Molecular Wire

An important function at the molecular level is photoinduced energy and electron transfer over long distances and/or along predetermined directions. This function can be obtained by linking donor and acceptor components by a rigid spacer, as illustrated in Fig. 14.

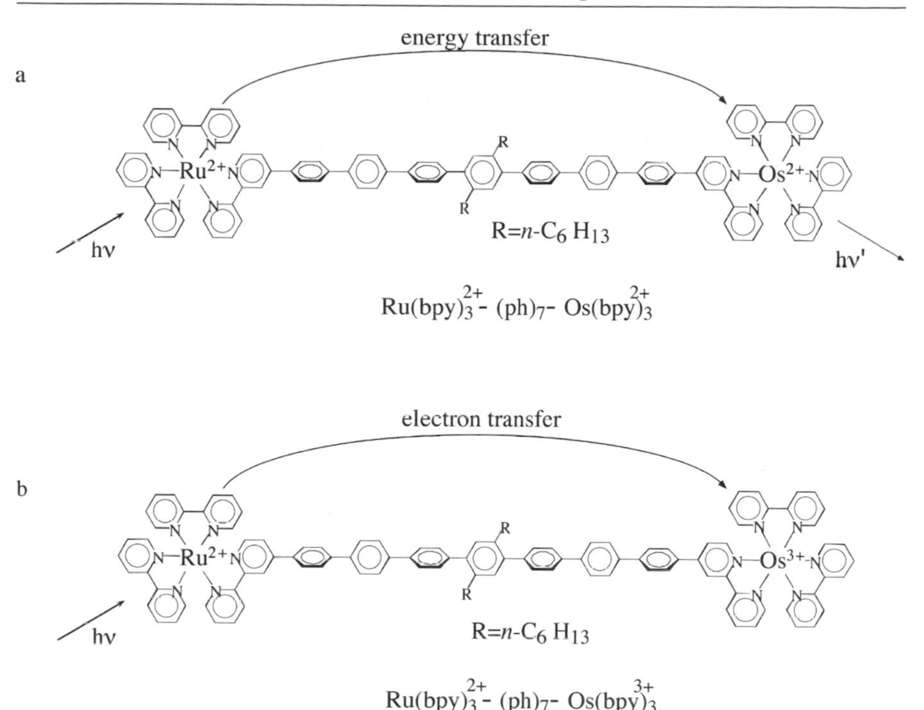

Fig. 14 Photoinduced energy (**a**) and electron (**b**) transfer processes in a molecular wire based on coordination compounds [76]

An example [76] is given by the $[Ru(bpy)_3]^{2+}$-$(ph)_n$-$[Os(bpy)_3]^{2+}$ compounds (bpy=2,2′-bipyridine; ph = 1,4-phenylene; $n = 3, 5, 7$), in which excitation of the $[Ru(bpy)_3]^{2+}$ unit is followed by electronic energy transfer to the ground state $[Os(bpy)_3]^{2+}$ unit, as shown by the sensitized emission of the latter. For the compound with $n = 7$ (Fig. 14a), the rate constant for energy transfer over the 4.2-nm metal-to-metal distance is $1.3 \times 10^6\ s^{-1}$. In the $[Ru(bpy)_3]^{2+}$-$(ph)_n$-$[Os(bpy)_3]^{3+}$ compounds, obtained by chemical oxidation of the Os-based moiety, photoexcitation of the $[Ru(bpy)_3]^{2+}$ unit causes the transfer of an electron to the Os-based one with a rate constant of $3.4 \times 10^7\ s^{-1}$ for $n = 7$ (Fig. 14b). Unless the electron added to the $[Os(bpy)_3]^{3+}$ unit is rapidly removed, a back electron-transfer reaction (rate constant $2.7 \times 10^5\ s^{-1}$ for $n = 7$) takes place from the $[Os(bpy)_3]^{2+}$ unit to the $[Ru(bpy)_3]^{3+}$ one.

Spacers with energy levels or redox states in between those of the donor and acceptor may help energy or electron transfer (hopping mechanism). Spacers whose energy or redox levels can be manipulated by an external stimulus can play the role of switches for the energy- or electron-transfer processes [48]. For a more thorough discussion of photoinduced energy- and electron-transfer processes in systems involving metal complexes, see [61].

5.2
An Antenna System

In suitably designed dendrimers, electronic energy transfer can be channeled toward a specific position of the array. Compounds of this kind play the role of antennas for light harvesting. We briefly illustrate an example involving luminescent lanthanide ions. For a more extended discussion of dendritic antenna systems, see [77].

Lanthanide ions are known to show a very long-lived luminescence which is a potentially useful property. Because of the forbidden nature of their electronic transitions, however, lanthanide ions exhibit very weak absorption bands, which is a severe drawback for applications based on luminescence. In order to overcome this difficulty, lanthanide ions are usually coordinated to ligands containing organic chromophores whose excitation, followed by energy transfer, causes the sensitized luminescence of the metal ion (antenna effect). Such a process can involve either direct energy transfer from the singlet excited state of the chromophoric group with quenching of the chromophore fluorescence [78], or, most frequently, via $S_1 \rightarrow T_1$ intersystem crossing followed by energy transfer from the T_1 excited state of the chromophoric unit to the lanthanide ion [79, 80].

Amide groups are known to be good ligands for lanthanide ions. The dendrimer shown in Fig. 15 is based on a benzene core branched in the 1, 3, and 5 positions, and it contains 18 amide groups in its branches and 24 chromophoric dansyl units in the periphery [81]. The dansyl units show strong absorption bands in the near-UV spectral region and an intense fluorescence band in the visible region. In acetonitrile/dichloromethane (5 : 1 v/v) solutions, the absorption spectrum and the fluorescence properties of the dendrimer are those expected for a species containing 24 noninteracting dansyl units. Upon addition of lanthanide ions to dendrimer solutions the following effects were observed [81]: (a) the fluorescence of the dansyl units is quenched; (b) the quenching effect is very large for Nd^{3+} and Eu^{3+}, moderate for Yb^{3+}, small for Tb^{3+}, and very small for Gd^{3+}; and (c) in the case of Nd^{3+}, Er^{3+}, and Yb^{3+} the quenching of the dansyl fluorescence is accompanied by the sensitized near-infrared emission of the lanthanide ion. Interpretation of the results obtained on the basis of the energy levels and redox potentials of the dendrimer and of the metal ions has led to the following conclusions: (1) at low metal ion concentrations, each dendrimer hosts only one metal ion; (2) when the hosted metal ion is Nd^{3+} or Eu^{3+}, all 24 dansyl units of the dendrimer are quenched with unitary efficiency; (3) quenching by Nd^{3+} takes place by direct energy transfer from the fluorescent (S_1) excited state of dansyl to a manifold of Nd^{3+} energy levels, followed by sensitized near-infrared emission from the metal ion ($\lambda_{max} = 1064$ nm for Nd^{3+}); (4) quenching by Eu^{3+} does not lead to any sensitized emission since a ligand-to-metal charge-transfer level lies be-

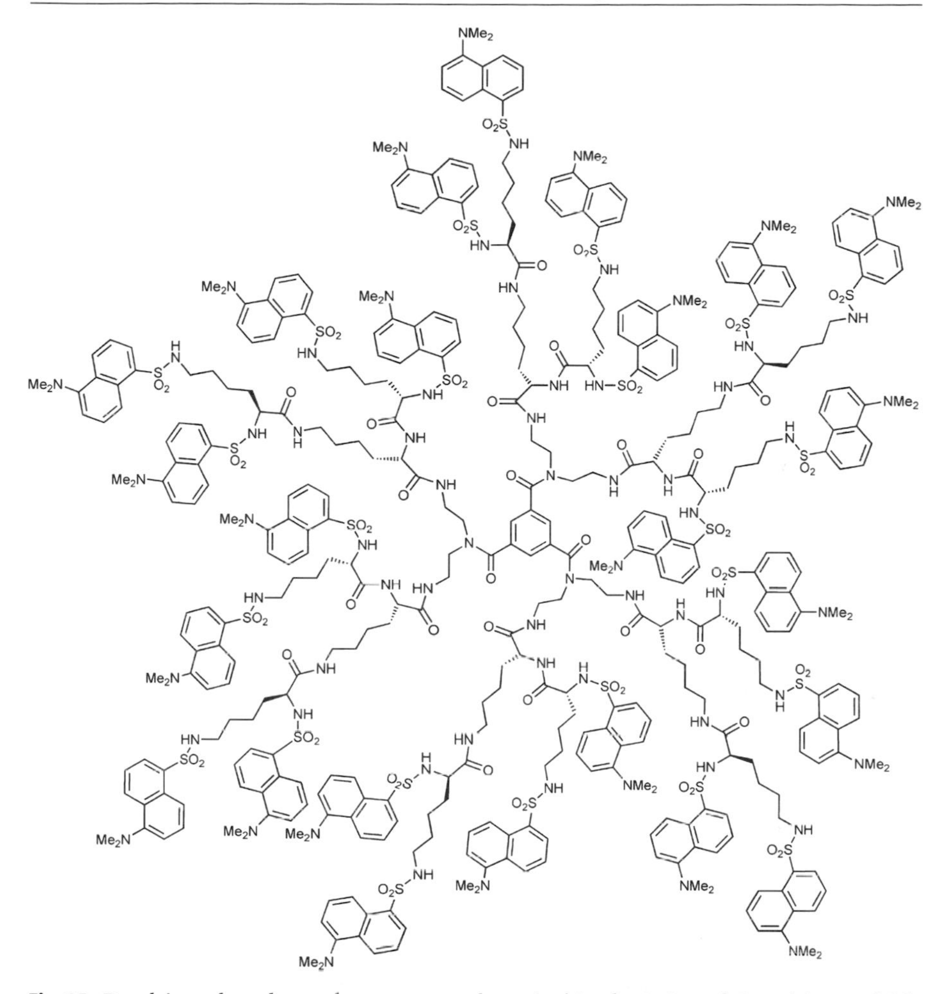

Fig. 15 Dendrimer based on a benzene core branched in the 1, 3, and 5 positions, which contains 18 amide groups in its branches and 24 chromophoric dansyl units in the periphery [81]

low the luminescent Eu^{3+} excited state; (5) in the case of Yb^{3+}, the sensitization of the near-infrared metal-centered emission occurs via the intermediate formation of an upper lying charge-transfer excited state; (6) the small quenching effect observed for Tb^{3+} is partly caused by a direct energy transfer from the fluorescent (S_1) excited state of dansyl; and (7) the very small quenching effect observed for Gd^{3+} is assigned to either induced intersystem crossing or, more likely, to charge perturbation of the S_1 dansyl excited state.

in acid or neutral aqueous solution causes the dissociation of a CN^- ligand from the metal coordination sphere ($\Phi = 0.31$), with a consequent increase in pH (Fig. 17b).

When an acid solution (pH=3.6) containing 2.5×10^{-5} mol L^{-1} **Ct** and 2.0×10^{-2} mol L^{-1} $[Co(CN)_6]^{3-}$ is irradiated at 365 nm most of the incident light is absorbed by **Ct**, which undergoes photoisomerization to **Cc**. Since the pH of the solution is sufficiently acid, **Cc** is rapidly protonated (Fig. 17a), with the consequent appearance of the absorption band with maximum at 434 nm and of the emission band with maximum at 530 nm characteristic of the AH^+ species. On continuing irradiation, it can be observed that such absorption and emission bands increase in intensity, reach a maximum value, and then decrease up to complete disappearance. In other words, AH^+ first forms and then disappears with increasing irradiation time. The reason for the off–on–off behavior of AH^+ under continuous light excitation is related to the effect of the $[Co(CN)_6]^{3-}$ photoreaction (Fig. 17b) on the **Ct** photoreaction (Fig. 17a). As **Ct** is consumed with formation of AH^+, an increasing fraction of the incident light is absorbed by $[Co(CN)_6]^{3-}$, whose photoreaction causes an increase in the pH of the solution. This change in pH not only prevents further formation of AH^+, which would imply protonation of the **Cc** molecules that continue to be formed by light excitation of **Ct**, but also causes the back reaction to **Cc** (and, then, to **Ct**) of the previously formed AH^+ molecules. Clearly, the examined solution performs like a threshold device as far as the input (light)/output (spectroscopic properties of AH^+) relationship is concerned.

Instead of a continuous light source, pulsed (flash) irradiation can be used [83]. Under the input of only one flash, a strong change in absorbance at 434 nm is observed, due to the formation of AH^+. After two flashes, however, the change in absorbance practically disappears. In other words, an output (434-nm absorption) can be obtained only when *either* input 1 (flash 1) *or* input 2 (flash 2) are used, whereas there is no output under the action of *none* or *both* inputs. This finding shows that the above-described system behaves according to XOR logic, under control of an intrinsic threshold mechanism (Fig. 17c).

Two important features of the above system should be emphasized: (1) intermolecular communication takes place in the form of H^+ ions, and (2) the input and output signals have the same nature (light) and the fluorescence output can be fed, in principle, into another device [85].

5.5 A Sunlight-Powered Nanomotor

In the last few years, a great number of light-driven molecular machines have been developed and the field has been extensively reviewed [48, 55, 86–92]. In several cases, the working principle of such machines exploits photoinduced electron transfer by $[Ru(bpy)_3]^{2+}$ or related coordination compounds.

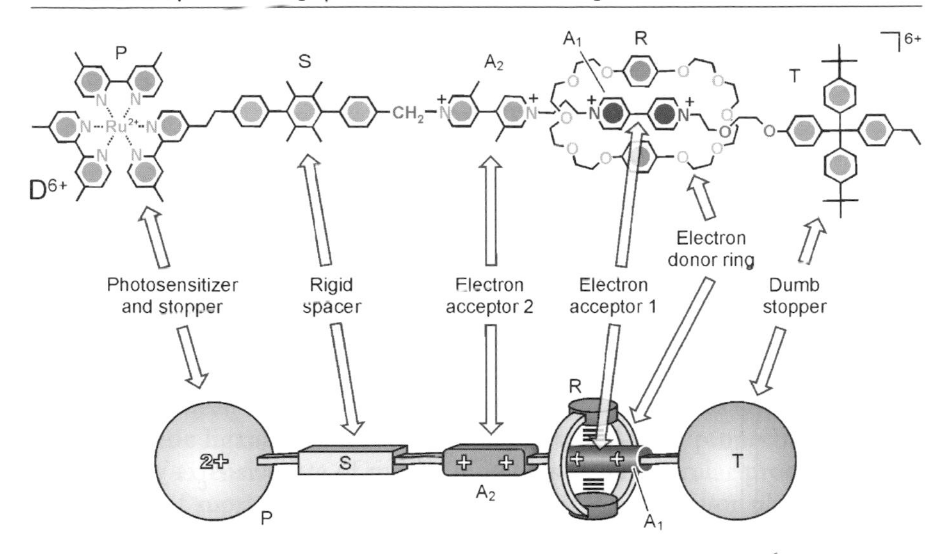

Fig. 18 Chemical formula and cartoon representation of the rotaxane $\mathbf{D}^{6+}$, showing its modular structure [94]

In order to achieve photoinduced ring switching in rotaxanes containing two different recognition sites in the dumbbell-shaped component, the thoroughly designed rotaxane ($\mathbf{D}^{6+}$) shown in Fig. 18 was synthesized [93–95]. This compound is made of the electron-donor macrocycle **R**, and a dumbbell-shaped component which contains (1) $[Ru(bpy)_3]^{2+}$ (**P**) as one of its stoppers, (2) a 4,4′-bipyridinium unit ($\mathbf{A}_1$) and a 3,3′-dimethyl-4,4′-bipyridinium unit ($\mathbf{A}_2$) as electron-accepting stations, (3) a *p*-terphenyl-type ring system as a rigid spacer (**S**), and (4) a tetraarylmethane group as the second stopper (**T**). The stable translational isomer is the one in which the **R** component encircles the $\mathbf{A}_1$ unit, in keeping with the fact that this station is a better electron acceptor than the other one. The strategy devised in order to obtain the photoinduced abacus-like movement of the **R** macrocycle, between the two stations $\mathbf{A}_1$ and $\mathbf{A}_2$ illustrated in Fig. 19, is based on the following four operations:

(a) *Destabilization of the stable translational isomer*: light excitation of the photoactive unit **P** (step 1) is followed by the transfer of an electron from the excited state to the $\mathbf{A}_1$ station, which is encircled by the ring **R** (step 2), with the consequent "deactivation" of this station; such a photoinduced electron-transfer process has to compete with the intrinsic excited-state decay (step 3).

(b) *Ring displacement*: the ring moves by Brownian motion from the reduced station $\mathbf{A}_1^-$ to $\mathbf{A}_2$ (step 4), a step that has to compete with the back electron-transfer process from $\mathbf{A}_1^-$ (still encircled by **R**) to the oxidized photoactive unit $\mathbf{P}^+$ (step 5).

87. Stoddart JF (ed) (2001) Special issue on molecular machines. Acc Chem Res 34:409
88. Venturi M, Credi A, Balzani V (2001) Electron-transfer processes in pseudorotaxanes. In: Balzani V (ed) Electron transfer in chemistry, vol 3. Wiley-VCH, Weinheim, p 501
89. Ballardini R, Gandolfi MT, Balzani V (2001) Electron-transfer processes in rotaxanes and catenanes In: Balzani V (ed) Electron transfer in chemistry, vol 3. Wiley-VCH, Weinheim, p 539
90. Ballardini R, Balzani V, Credi A, Gandolfi MT, Venturi M (2001) Acc Chem Res 34:445–455
91. Sauvage JP (ed) (2001) Molecular machines and motors. Springer, Berlin
92. Kelly TR (ed) (2005) Molecular machines. Top Curr Chem, vol 262. Springer, Berlin
93. Ashton PR, Ballardini R, Balzani V, Credi A, Dress R, Ishow E, Kleverlaan CJ, Kocian O, Preece JA, Spencer N, Stoddart JF, Venturi M, Wenger S (2000) Chem Eur J 6:3558
94. Balzani V, Credi A, Ferrer B, Silvi S, Venturi M (2005) In: Kelly TR (ed) Molecular machines. Top Curr Chem 262:1
95. Balzani V, Clemente-León M, Credi A, Ferrer B, Venturi M, Flood AH, Stoddart JF (2006) Proc Natl Acad Sci USA 103:1178

Top Curr Chem (2007) 280: 37–67
DOI 10.1007/128_2007_141

Published online: 11 July 2007

Photochemistry and Photophysics of Coordination Compounds: Chromium

Noel A. P. Kane-Maguire

Department of Chemistry, Furman University, 3300 Poinsett Highway, Greenville, SC 29613, USA
noel.kane-maguire@furman.edu

Abstract The study of the photochemistry and photophysics of octahedral and pseudo-octahedral Cr(III) complexes has a rich history. An initial discussion is devoted to a general appraisal of the state of these two subjects up to December 1998, after providing a framework of state energy levels and radiative and non-radiative relaxation processes relevant to Cr(III) systems. The remaining sections cover some of the more active areas in the Cr(III) field, such as ultrafast dynamics, photosubstitution, thermally activated

The dominant ligand field bands in the UV-visible absorption spectra are associated with the $^4A_{2g} \rightarrow {}^4T_{2g}$ and $^4A_{2g} \rightarrow {}^4T_{1g}$ transitions, since the corresponding absorptions generating the two doublet excited states are both Laporte and spin multiplicity forbidden. Photochemical and photophysical studies of Cr(III) species are, therefore, usually restricted to initial excitation into one of the quartet excited states. The spin-allowed absorption bands for the $^4A_{2g} \rightarrow {}^4T_{2g}$ and $^4A_{2g} \rightarrow {}^4T_{1g}$ transitions are broad. This is a consequence of the $^4T_{2g}$ and $^4T_{1g}$ excited states both having an electron residing in an e_g antibonding σ^* orbital (Fig. 1), which results in a large nuclear displacement relative to the ground state. For excitation into the higher lying $^4T_{1g}$ level, very fast internal conversion (IC) occurs to the $^4T_{2g}$ state with near unit efficiency. Under normal photochemical conditions (i.e., in solution near room temperature) $^4T_{2g} \rightarrow {}^4A_{2g}$ fluorescence is rarely observed [12, 13], due to $^4T_{2g} \rightarrow {}^2E_g$ intersystem crossing (ISC) being an unusually rapid process [4, 13–15] and often occurring with high efficiency (see Table 4 in [4]). With only occasional exceptions, Cr(III) complexes in rigid low temperature media exhibit intense, long-lived phosphorescence from the ^2E_g level generated by ISC. Very little geometric change is expected between the $^4A_{2g}$ ground state and ^2E_g excited state, due their common $(t_{2g})^3$ orbital parentage. As a result, low temperature $^2E_g \rightarrow {}^4A_{2g}$ phosphorescence spectra often display sharp, highly resolved fine structure, which has led to a very extensive literature (including medium and temperature effects) on the photophysical properties of these systems [2, 4, 9, 11, 16].

In room temperature (rt) solution, $^4A_{2g} \rightarrow {}^4T_{2g}$ (or $^4A_{2g} \rightarrow {}^4T_{1g}$) excitation often leads to facile substitution of one or more bound ligands by solvent or an added nucleophile [1–4]. This observation does not, however, preclude the possibility of reaction out of the lower lying ^2E_g level, since this state is subsequently populated by rapid and efficient $^4T_{2g} \rightarrow {}^2E_g$ ISC. An enormous effort has been expended over the last 30 years in an attempt to determine the relative photochemical roles of the $^4T_{2g}$ and ^2E_g excited states. Importantly, a significant number of Cr(III) complexes exhibit relatively long-lived ($\geq$ 100 ns) phosphorescence in solution near rt, and this emission can be bimolecularly quenched by added reagents. Comparing photoreaction data for experiments carried out in the presence and absence of these added reagents (an approach first pioneered by Chen and Porter [17]) has in many instances proved very informative.

The most definitive quenching studies have been for the strong field hexacyano complex $[Cr(CN)_6]^{3-}$ [18] and the pentacyano species $[Cr(CN)_5(X)]^{n-}$, where X = NH_3 [19], pyridine (py) [20], and NCS^- [5]. Under experimental conditions of total emission quenching, no reaction quenching was detected. Such data provide compelling evidence for the reaction proceeding exclusively out of the $^4T_{2g}$ level (a similar conclusion was reached for $[Cr(CN)_6]^{3-}$ based on sensitization studies [21, 22]). For most Cr(III) systems, however, total phosphorescence quenching is accompanied by significant but less than total reaction quenching. Definitively establishing the precise role of the doublet level in this quenched reaction component has proven an elusive goal, with competing options including direct ^2E_g reaction, ^2E_g tunneling to a ground state intermediate (GSI) surface, or "delayed" quartet excited state reaction via thermally activated $^2E_g \rightarrow {}^4T_{2g}$ back-intersystem crossing, BISC (Fig. 2).

This debate has been exhaustively discussed elsewhere [4, 6, 23, 24], and will not be a focus of this review. From the outset, it was appreciated that the $^4T_{2g}$ level of $(t_{2g})^2(e_g)^1$ orbital parentage was an attractive candidate for substitution chemistry, based on the occupation of an e_g orbital which is σ^* antibonding with respect to the metal–ligand (M – L) bond. At present, the most widely employed theoretical model for rationalizing Cr(III) photosubstitution behavior, assuming quartet reactivity, is the semi-empirical symmetry restricted angular overlap model (AOM) developed by Vanquickenborne and Ceulemans [25–27]. For mixed ligand systems it has had considerable success in predicting relative ligand labilities based on identifying the plane of labilization and assuming that the ligand with the smallest excited state M – L bond strength is preferentially substituted. A further strength of the model is the rationalization it provides for the stereochemical change that is a common feature of Cr(III) photochemistry (in contrast to their corresponding thermal behavior), especially for cases of axial ligand loss in mixed ligand systems of D_{4h} or C_{4v} symmetry [4, 28]. A more recent ab initio study of the photochemistry of Cr(III) ammine systems yielded results in good agreement with the earlier AOM calculations [29]. Tris-polypyridyl Cr(III) complexes may prove to be an exception to this apparent preference for quartet excited state reactivity. Early quenching studies on $[Cr(phen)_3]^{3+}$ (where phen = 1,10-phenanthroline) revealed up to 95% reaction quenching in the presence of doublet quenchers such as I^- and NCS^- [30, 31], and the data from subsequent and more detailed investigations were most readily interpreted in terms of a direct doublet excited state reaction pathway [32–34].

The remaining sections of this chapter are devoted to a discussion of developments since December 1998 in a range of different focus areas of Cr(III) photochemistry and photophysics. For convenience, the state term symbols for O_h symmetry shown in Fig. 2 are usually employed during these discussions.

3
Ultrafast Dynamics of Cr(III) Ligand Field Excited States

Ultrafast time-resolved absorption spectroscopy constitutes one of the most exciting and promising new frontiers in transition metal photochemistry and photophysics. The term ultrafast is applied to photoprocesses that occur on

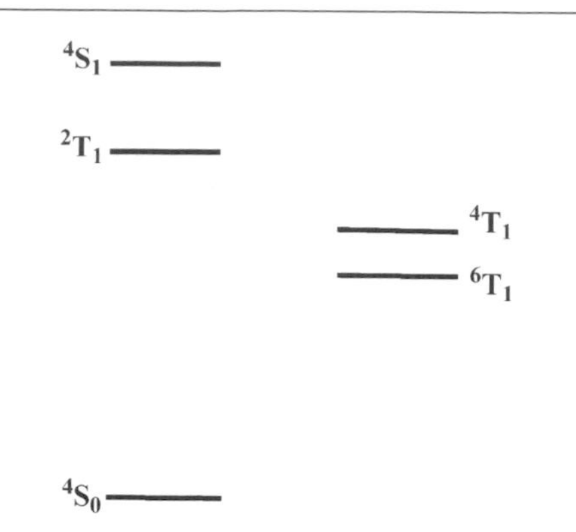

Fig. 7 State energy level diagram for [Cr(porphyrin)(Cl)(L)] complexes, showing the porphyrin (π,π^*) levels following weak coupling with the *d* orbitals of the paramagnetic Cr(III) center

tation of [Cr(porphyrin)(Cl)(L)] into the $^4S_1$ ($\pi-\pi^*$) excited state, confirmed the formation of the five-coordinate complex, [Cr(porphyrin)(Cl)], produced by the photodissociation of the axial ligand L. Spectral evidence was also found for generation of the thermally equilibrated $^4T_1$ and $^6T_1$ excited states.

The quantum yield, ϕ_{diss}, for the photodissociation of L from [Cr(porphyrin)(Cl)(L)]^0 was found to asymptotically decrease with increasing dissolved O_2 concentrations towards a constant value. This suggested the presence of a quenchable dissociation pathway attributed to the longer-lived $^4T_1$ and $^6T_1$ levels, and a non-quenchable reaction component associated with the short-lived ($\ll 1$ ns) $^4S_1$ level. The ϕ_{diss} values were also found to vary markedly with the porphyrin ring and axial ligands present. Figure 8 summarizes the electronic energy dissipation processes proposed for these Cr(III) porphyrin systems.

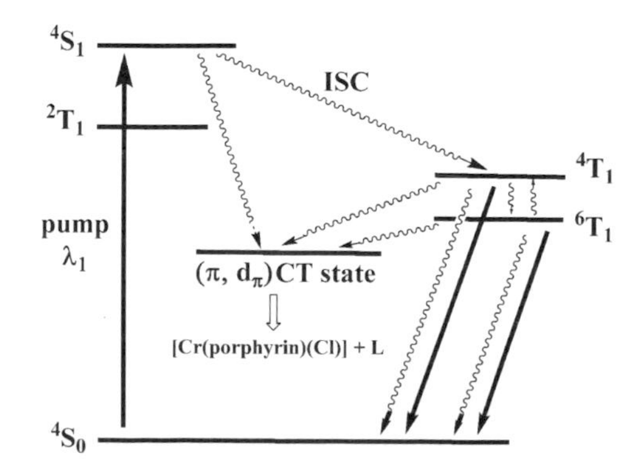

Fig. 8 State energy level diagram for [Cr(porphyrin)(Cl)(L)] complexes showing the principal relaxation processes following $^4S_0 \rightarrow {}^4S_1$ excitation. *Full arrows* represent radiative processes, whereas *wavy arrows* refer to radiationless decay pathways

As shown in Fig. 8, the quenchable and non-quenchable dissociation pathways are both thought to proceed through lower-lying porphyrin π–Cr d_π charge transfer (CT) states [48]. For example, an increase in the electron density in a Cr d_π orbital with a *z*-axis component should lead to a weakening in the Cr-axial ligand bond. The rich photobehavior of these systems suggests that they would make good candidates for future ultrafast spectroscopic studies.

5 Thermally Activated ^2E_g Excited State Relaxation Studies of Sterically Constrained Systems

Early studies on Cr(III) complexes of the type $[Cr(N_4)X_2]^{n+}$ where N_4 is the macrocyclic ligand cyclam or tet a (see Fig. 9) and X = Cl^- [53, 54], CN^- [55, 56], NH_3 [56–58], and F^- [59] revealed a marked difference in the photobehavior of the geometric isomers. All the *trans* isomers are photoinert while the *cis* species are photoreactive. The cyano, ammine, and fluoro systems drew particular attention because of their strong emission in rt solution, with accompanying lifetimes almost identical to those reported at 77 K. These observations were attributed to the steric rigidity of the macrocyclic ring restricting access to the thermally activated photochemical relaxation channels available to their non-macrocylic analogs.

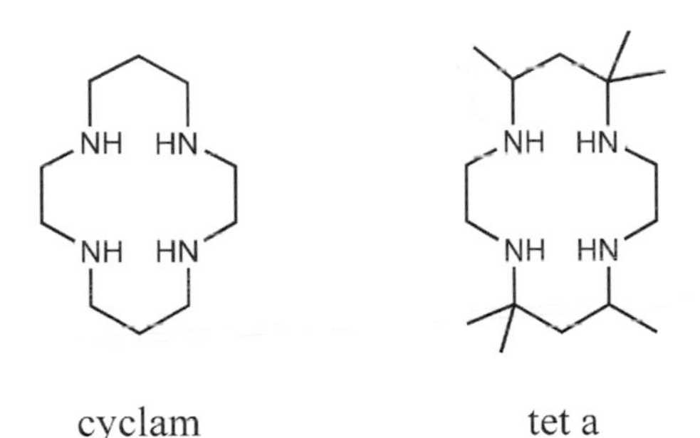

Fig. 9 Macrocyclic-N_4 ligands, cyclam and tet a

An extensive literature now exists on the effects of ligand steric constraint on ^2E_g excited state relaxation [4, 60–71], with studies by Endicott and coworkers being especially noteworthy. Hexaam(m)ine Cr(III) systems have been one key area of study of Endicott's group, where a range of complexes were synthesized containing ligands that would be trigonally strained if coordinated octahedrally to Cr(III). Their studies provided convincing evidence that the more trigonally strained ligand systems underwent more rapid ^2E_g deactivation. In an illustrative example [63], the photobehavior of $[Cr(en)_3]^{3+}$ was compared with that of the quasi-cage complex $[Cr(sen)]^{3+}$, where $[Cr(sen)]^{3+}$ can be regarded as a $[Cr(en)_3]^{3+}$ analog with a neopentyl cap bonded in a facial position.

X-ray crystallographic data revealed that the CrN_6 microsymmetry is virtually identical for these two complexes, with the NCrN bond angles in $[Cr(sen)]^{3+}$ being slightly closer to octahedral. These data and MM2 calculations also established considerable trigonal strain in the neopentyl cap for $[Cr(sen)]^{3+}$.

Both compounds were found to have similar $^2E_g \rightarrow {}^4A_{2g}$ emission lifetimes at 77 K (120 μs and 171 μs for the en and sen complexes, respectively), and fairly comparable quantum yields for photoaquation in rt solution (0.27 and 0.10, respectively). However, the ^2E_g lifetime for $[Cr(sen)]^{3+}$ in ambient solution was a factor of 10^4 times shorter than that for $[Cr(en)_3]^{3+}$. The authors attributed this difference to a thermally activated ^2E_g deactivation channel promoted by steric factors associated with the sen complex. The general conclusion from this body of work was that large amplitude trigonal twists can facilitate thermally activated ^2E_g relaxation for a range of sterically constrained hexaam(m)ine Cr(III) complexes [64]. The authors also suggest that this relaxation pathway may have mechanistic implications for the photoracemization of Cr(III) species with D_3 symmetry [63]. Such a reaction channel could, for example, facilitate the trigonal twist pathway invoked for the observed photoinversion of Λ-*fac*-[Cr(S-trp)$_3$] to Δ-*fac*-[Cr(S-trp)$_3$] (where *S*-trp is the bidentate amino acid ligand *S*-tryptophan) [72].

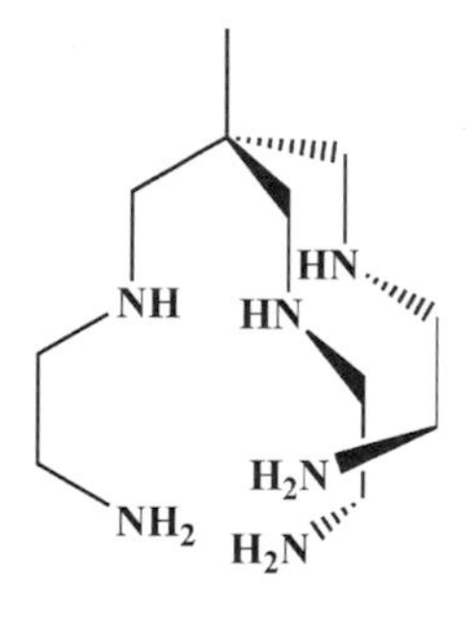

Fig. 10 Quasi-cage N_6 ligand, sen

The earlier studies on macrocyclic *cis*- and *trans*-$[Cr(N_4)X_2]^{n+}$ complexes (where X is NH_3 or CN^-) were also expanded by Endicott's group to include systems where stereochemical perturbations were introduced by the presence of methyl substituents in the macrocylic ring in positions near the Cr–X coordination sites [65]. Their analysis of X-ray data and MM2 calculations supported the hypothesis that the more facile thermally activated ^2E_g relaxation of the *cis*-$[Cr(N_4)X_2]^{n+}$ systems is predominantly a stereochemical effect. It was also argued that ${}^2E_g$ back-intersystem crossing (BISC) was not likely to be an important component of ${}^2E_g$ excited state deactivation for Cr(III) complexes with N_6 or N_4C_2 chromophores [64]. This latter conclusion has been questioned by Kirk [4,68], and in Sect. 5.1 the subject is discussed further. In Sect. 5.2 some very recent work by the Wagenknecht group employing a new series of sterically constrained N_4-macrocycles is featured [69–71].

5.1
$[Cr(sen)]^{3+}$ and $[Cr[18]aneN_6]^{3+}$

5.1.1
$[Cr(sen)_3]^{3+}$

In a recent report, Kirk and coworkers have reinvestigated the photobehavior of $[Cr(sen)]^{3+}$, and compared it with that for the macrocyclic complex, $[Cr[18]aneN_6]^{3+}$ [68].

The data obtained for $[Cr(sen)]^{3+}$ supported the earlier report of a very short doublet lifetime in rt aqueous solution, and the photoaquation quantum yield of 0.10 determined upon ${}^4A_{2g} \rightarrow {}^4T_{2g}$ excitation was in excellent agreement with that recorded earlier.

However, based on a more detailed investigation of $Cr(sen)]^{3+}$ photoaquation, it was proposed that this process occurs via the ${}^4T_{2g}$ excited state after back-intersystem crossing (BISC). The more convincing argument presented was that direct irradiation into the ${}^2E_g$ state yielded a photoaquation quantum yield 22% lower than that for ${}^4T_{2g}$ excitation excitation. However, as noted by the authors, direct spin-forbidden doublet excitation experiments are fraught with difficulties. Their second argument was based on a deter-

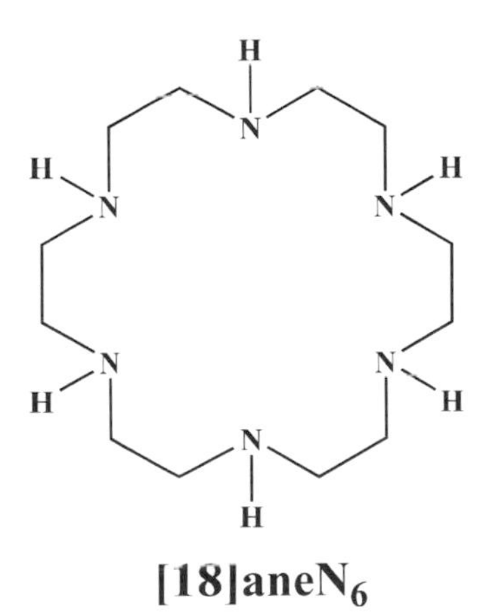

Fig. 11 Macrocyclic-N_6 ligand, $[18]aneN_6$

cells are hypoxic [111], the increase in photodamage at lower O_2 levels may provide a selectivity advantage for these $[Cr(diimine)_3]^{3+}$ reagents in terms of their future potential as phototherapeutic agents.

In another aspect of this study [118], a mathematical analysis of the emission quenching data was undertaken. Representative steady-state intensity and lifetime Stern–Volmer (SV) plots for quenching of $[Cr(phen)_3]^{3+}$ emission by calf thymus B-DNA are shown in Fig. 18. From the lifetime data, a bimolecular quenching rate constant of $1.1 \times 10^8\ M^{-1}\ s^{-1}$ was extracted (a value close to that anticipated for a diffusion controlled process). In contrast, the steady-state SV plot showed strong upward curvature at higher DNA concentrations. This observation was attributed to the formation of a non-luminescent $[Cr(phen)_3]^{3+}$/DNA ion pair, and allowed an estimation to be made for the binding constant with DNA ($K_{DNA} \approx 4000\ M^{-1}$).

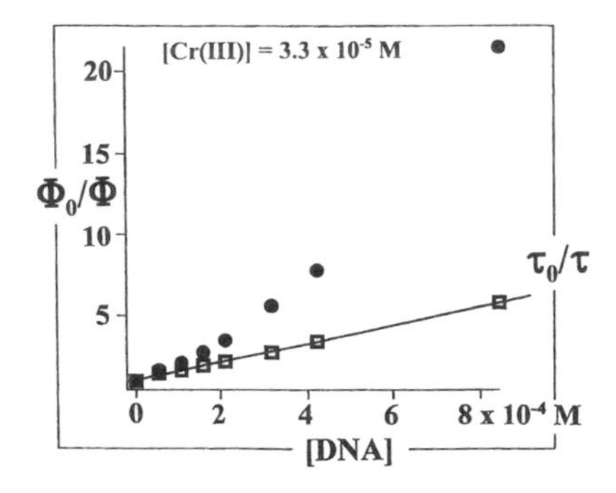

Fig. 18 Stern–Volmer plots for $[Cr(phen)_3]^{3+}$ emission quenching in air-saturated 50 mM Tris-HCl buffer (pH 7.4) by calf thymus B-DNA at 22 °C: • steady-state data, □ lifetime data

A limitation of this initial work with $[Cr(bpy)_3]^{3+}$ and $[Cr(phen)_3]^{3+}$ is the relatively small binding constants of these compounds with DNA. In a subsequent study [119], the photoredox behavior of the complex $[Cr(phen)_2(DPPZ)]^{3+}$ with DNA was investigated, where the third diimine ligand is dipyridophenazine, DPPZ. The value of K_{DNA} increased by two orders of magnitude, consistent with the known ability of the DPPZ ligand to intercalate into DNA base stacks [113]. Perhaps more importantly, the complex was found to have an $E^o(^*Cr^{3+}/Cr^{2+})$ value 80 mV more positive than that for $[Cr(phen)_3]^{3+}$, which placed it in the thermodynamic threshold range required for direct oxidation of the nucleobase adenine [115]. In accord with this thermodynamic argument, SV plots of the quenching of the emission lifetime of $[Cr(phen)_2(DPPZ)]^{3+}$ in the presence of deoxyguanosine-5′-monophosphate and deoxyadenosine-5′-monophosphate yielded quenching rate constants of $2.4 \times 10^9\ M^{-1}\ s^{-1}$ and $1.8 \times 10^7\ M^{-1}\ s^{-1}$, respectively [119].

More recently [120], a report by Vaidyanathan and Nair describes nucleobase photooxidation by the terpyridine Cr(III) derivatives $[Cr(ttpy)_2]^{3+}$

and $[Cr(Brphtpy)_2]^{3+}$ (where ttpy = *p*-tolylterpyridine and Brphtpy = *p*-bromophenylterpyridine). The two complexes were reported to emit strongly in rt aqueous solution, although no emission lifetimes or spectra (except wavelength maxima) were provided. Such emission is quite remarkable in view of the exceedingly weak emission and very short lifetime ($\approx 0.05\ \mu s$) found for the parent terpyridine complex, $[Cr(tpy)_2]^{3+}$ [103]. Based on CV data and the reported emission spectral maxima, exceptionally high values for $E^o(^*Cr^{3+}/Cr^{2+})$ were assessed for $[Cr(ttpy)_2]^{3+}$ and $[Cr(Brphtpy)_2]^{3+}$ (1.65 V and 1.85 V, respectively). Consistent with these thermodynamic observations, both complexes were demonstrated to be very powerful photooxidants. This was especially true for $[Cr(Brphtpy)_2]^{3+}$, where its emission was quenched by all four mononucleotides (including deoxythymidine-5′-monophosphate). This statement, however, requires that the labels for Figs. 4A and B in the paper were accidentally reversed.

8
Photoredox Involving Coordinated Ligands

Whereas Sect. 7 was concerned with examples of intermolecular electron transfer between Cr(III) excited states and external substrates, attention is directed in the present section to cases of intramolecular redox chemistry involving the coordinated ligands. These studies have usually involved photoexcitation into high-energy LMCT excited states involving the ligand in question, which often results in the transient formation of a Cr(II)/ligand radical pair. The subject has been reviewed by Kirk [4], and some representative examples of molecules previously investigated are $[Cr(NH_3)_5Br]^{2+}$ [121] and *trans*-$[Cr(tfa)_3]$ (where tfa is the anion of 1,1,1-trifluoro-2,4-pentanedione) [122]. Some of the more recent contributions in this area are discussed in the following two sections.

8.1
Photolabilization of NO from Cr(III)-Coordinated Nitrite

It has been recently established that nitric oxide (NO) regulates a number of mammalian biological processes, including blood pressure, neurotransmission, and smooth muscle relaxation [123]. Additionally, tumor cells are particularly sensitive to NO, which induces programmed cell death [124] and limits metasis [125]. In response to these findings, Ford and coworkers have developed a range of air-stable, water-soluble nitrito-Cr(III) macrocyclic complexes, which display photochemically activated NO release [126–128]. The initial study involved the complex *trans*-$[Cr(cyclam)(ONO)_2]^+$ [126], which for convenience is labeled structure **I** in Fig. 19.

Top Curr Chem (2007) 280: 69–115
DOI 10.1007/128_2007_128

Published online: 24 May 2007

Photochemistry and Photophysics of Coordination Compounds: Copper

Nicola Armaroli (✉) · Gianluca Accorsi · François Cardinali · Andrea Listorti

Molecular Photoscience Group, Istituto per la Sintesi Organica e la Fotoreattività, Consiglio Nazionale delle Ricerche, Via Gobetti 101, 40129 Bologna, Italy
nicola.armaroli@isof.cnr.it

Abstract Cu(I) complexes and clusters are the largest class of compounds of relevant photochemical and photophysical interest based on a relatively abundant metal element. Interestingly, Nature has given an essential role to copper compounds in some biological systems, relying on their kinetic lability and versatile coordination environment. Some basic properties of Cu(I) and Cu(II) such as their coordination geometries and electronic levels are compared, pointing out the limited significance of Cu(II) compounds (d^9 configuration) in terms of photophysical properties. Well-established synthetic protocols are available to build up a variety of molecular and supramolecular architectures can be classified in three main categories, i.e. anionic complexes (e.g. alo-complexes), neutral clusters and cationic complexes. The photochemistry of Cu(I) complexes, also related to environmental aspects, has already been reviewed [3, 4], here we will essentially focus on photophysics. Anionic complexes do not exhibit attractive photophysical properties (e.g. luminescence), unlike cluster and cationic complexes which show a very rich photophysical behavior. Among the latter, the most extensively investigated are NN-type (where NN indicates a chelating imine ligand, typically 1,10-phenanthroline) or PP-type (where PP denotes a bisphosphine ligand). Both homoleptic $[Cu(NN)_2]^+$ and heteroleptic $[Cu(NN)(PP)]^+$ motifs have been investigated.

The coordination behavior of Cu(I) is strictly related to its electronic configuration. The complete filling of *d* orbitals (d^{10} configuration) leads to a symmetric localization of the electronic charge. This situation favors a tetrahedral disposition of the ligands around the metal center in order to locate the coordinative sites far from one another and minimize electrostatic repulsions (Fig. 1). Clearly, the complete filling of *d* orbitals prevents *d*-*d* metal-centered electronic transitions in Cu(I) compounds. On the contrary, such transitions are exhibited by d^9 Cu(II) complexes and cause relatively intense absorption bands in the visible (VIS) spectral window. The lowest ones extend into the near infrared (NIR) region (above 800 nm for the Cu(II) aqua ion) [5] and deactivate via ultrafast non-radiative processes. The fact that the lowest electronic states of Cu(II) complexes are ultra-short lived make them far less interesting than Cu(I) complexes from the photophysical point of view.

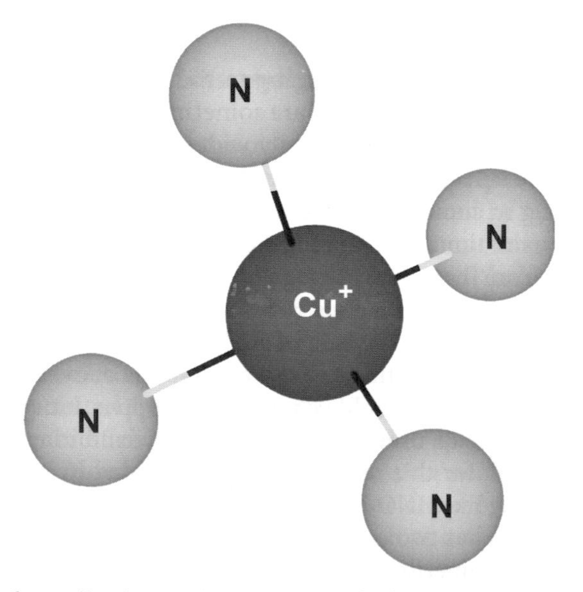

Fig. 1 Tetrahedral coordination environment typical of Cu(I) complexes

Cu(I) cluster compounds are characterized by a variety of emitting electronic levels whereas Cu(I) cationic complexes show only luminescence originating from metal-to-ligand charge-transfer (MLCT) states, as long as empty π orbitals are easily accessible in the ligands. Such MLCT transitions, which clearly take advantage of the low oxidation potential of Cu(I), are also commonly observed in other classes of coordination compounds, for example those of d^6 metals like Ru(II)-bipyridines [6] and Ir(III)-phenylpyridine complexes [7].

MLCT electronic transitions in coordination compounds are normally more intense when compared to MC (metal-centered) ones since they do not undergo the same prohibitions by orbital symmetry; accordingly MLCT absorption bands exhibit relatively high molar extinction coefficients. As far as emission is concerned, when MLCT excited states are the lowest-lying, they are generally characterized by long lifetimes, and potentially intense luminescence, even though exceptions are possible (vide infra). Complexes exhibiting long-lived MLCT excited states have been extensively investigated in the last decades both for a better comprehension of fundamental phenomena [8, 9] and for potential applications related to solar light harvesting and conversion [10–12]. Among them the highest attention was probably devoted to Ru(II) [13], Os(II) [14] and, more recently, Ir(III) [7] complexes, however, economical and environmental considerations make Cu(I) compounds interesting alternatives [15].

As extensively discussed in the literature, long-lived luminescent MLCT excited states of d^6 metal complexes, in particular those of Ru(II), can be strongly affected by the presence of upper lying MC levels. The latter can be partially populated through thermal activation from the MLCT states and prompt non-radiative deactivation pathways and photochemical degradation [6, 16]. Closed shell d^{10} copper(I) complexes cannot suffer these kinds of problems, but undesired non-radiative deactivation channels of their MLCT levels can be favored by other factors, as will be discussed in detail further on in this review. An orbital diagram illustrating the electronic transitions of Ru(II) and Cu(I) complexes is reported in Fig. 2.

1.3
Copper in Biology

Copper, even if present in traces, is an essential metal for the growth and development of biological systems. Copper plays a fundamental role in cerebral activity, nervous and cardiovascular systems, oxygen transport and cell protection against oxidation. Copper is important to strengthen the bones and to guarantee the performances of the immune system [17].

Metals are commonly found as natural constituents of proteins and, in the course of evolution, Nature has learned how to use the special properties of metal ions to perform a wide variety of specific functions associated

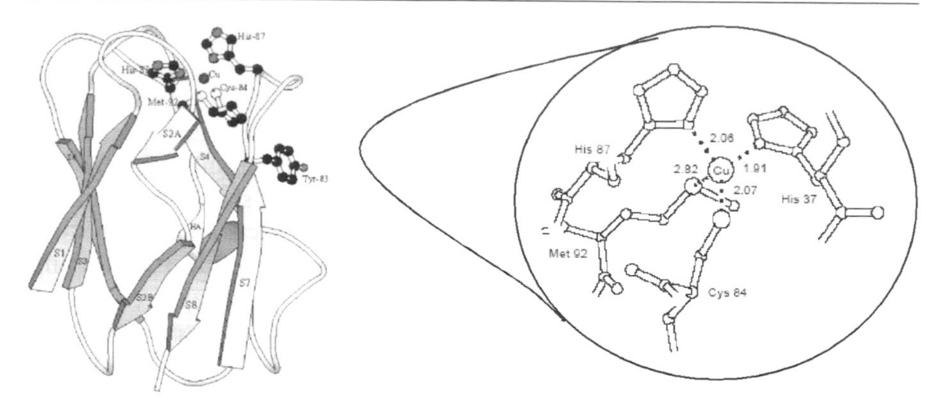

Fig. 3 The blue copper site inside plastocyanin

sitions of normal tetragonal Cu(II) complexes with $\varepsilon \approx 40\,M^{-1}cm^{-1}$ at ca. $16\,000\,cm^{-1}$ (≈ 620 nm), the blue copper site has an intense absorption band at $16\,000\,cm^{-1}$ with $\varepsilon \approx 5000\,M^{-1}cm^{-1}$ Fig. 4 [20]. This result is a consequence of an inversion of the ligand-to-metal charge transfer (LMCT) pattern for the blue copper site that rises from its particular ligands distribution. As can be seen in Fig. 5 variation of the typical overlapping between the orbitals of copper and those of the ligands leads to an inversion of the relative absorption intensity, the final result is an enhancement of the absorption on the low energy side [21].

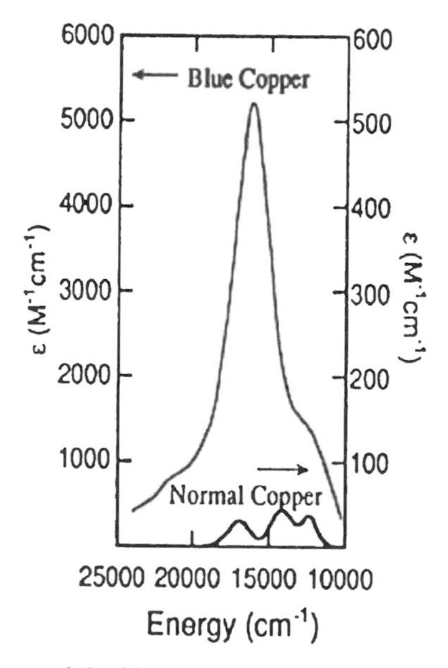

Fig. 4 Absorption spectrum of the blue copper site in plastocyanin (Reprinted from [20] with permission, © (2006) American Chemical Society)

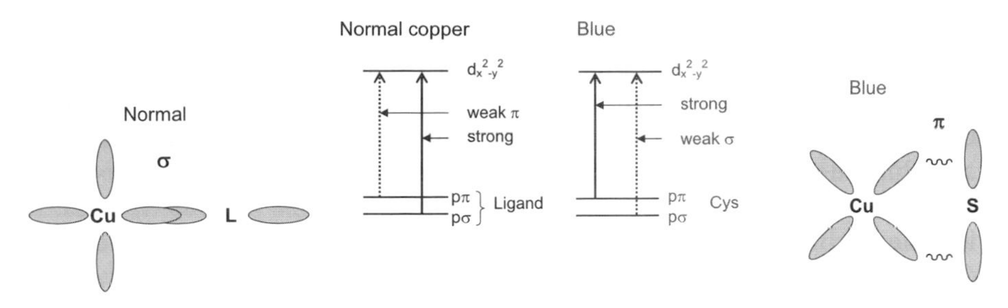

Fig. 5 Inverted intensity pattern of ligand-to-metal charge transfer absorption transitions for the blue copper site compared to a regular Cu(II) complex. L = generic organic ligand; S = sulfur coordinating site of a cysteine residue

In photosynthesis, plastocyanin functions as an electron transfer relay between cytochrome f (inside cytochrome b_6f complex) and $P700^+$. Cytochrome b_6f complex (from photosystem II) and $P700^+$ (from photosystem I) are both membrane-bound proteins with exposed residues on the lumen-side of the thylakoid membrane of chloroplasts. Cytochrome f acts as an electron donor while P700+ accepts electrons from reduced plastocyanin (Fig. 6) [18].

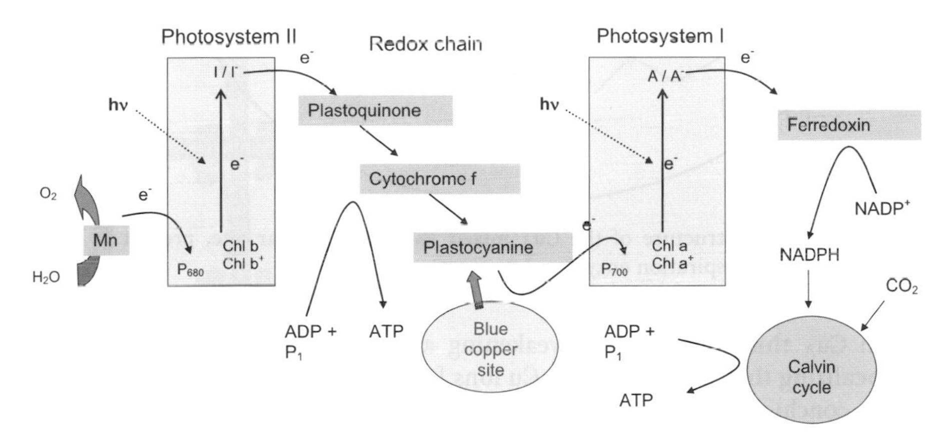

Fig. 6 The so-called Z-scheme of photosynthesis

Plastocyanin (Cu^{2+}Pc) is reduced by cytochrome f to Cu^+Pc which eventually diffuses through the lumen until recognition/binding occurs with $P700^+$, which oxidizes Cu^+Pc back to Cu^{2+}Pc. The electronic structure of the blue copper is crucial for an efficient electron transfer in which Cu(II) is reduced to Cu(I). The tetrahedral organization of the Cu(II) site minimizes the reorganizational energy λ increasing the rate of the process, according to Marcus theory [22].

pseudo-rotaxanes [34, 35], knots [36, 37], dendrimers [38, 39], helices [40–42], polynuclear hosts [43] etc. as originally developed by Sauvage, Dietrich-Buchecker and coworkers [44]. Some of these fascinating structures are depicted in Fig. 8.

Most notably, some suitably engineered supramolecular architectures based on $[Cu(NN)_2]^+$ cores are able to carry out motion at the molecular level upon chemical [45], or electrochemical/photochemical stimulation [32, 46] behaving as molecular machine prototypes [47, 48]. For instance a [2]-catenate made of two different rings, one with a phenanthroline fragment the other bearing both a phenanthroline and a terpyridine unit, undergoes spontaneous and reversible molecular rearrangements (Fig. 9 steps (B) and (D)) upon oxidation (step A) and subsequent reduction (step C). Rearrangements are driven by the different preferential coordination geometries of Cu(I) and Cu(II), i.e. tetra- vs. pentacoordination [46].

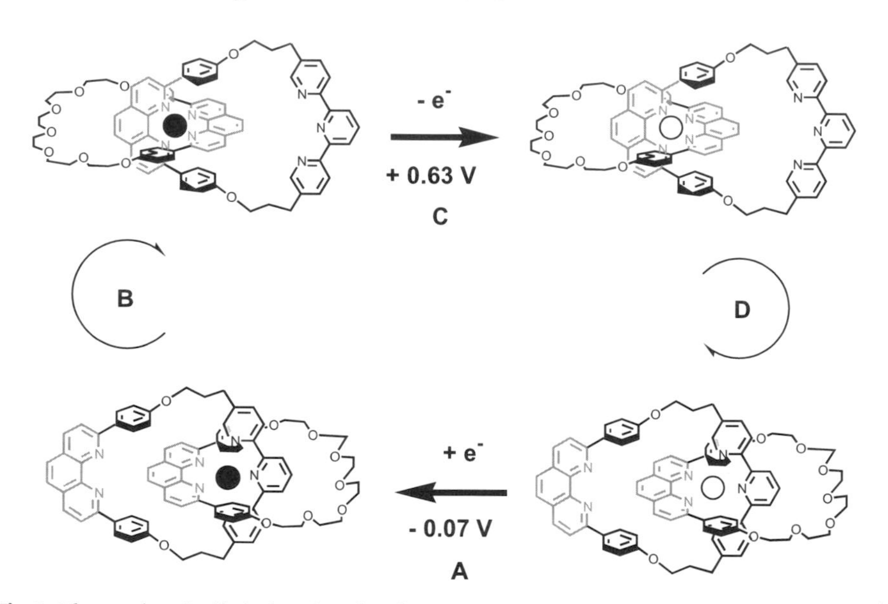

Fig. 9 Electrochemically induced molecular motions in a catenane containing a $[Cu(NN)_2]^+$ center and a free tpy ligand. The spontaneous motion is driven by the different preferential coordination geometry of Cu(I) vs. Cu(II)

More recently, Schmittel and coworkers have made new supramolecular systems based on $[Cu(NN)_2]^+$-type building blocks such as racks [49], grids [50], boxes [51], and macrocycles [52]. Control of the sophisticated heteroleptic architectures has been achieved by exploiting the HETPHEN (HETeroleptic bisPHENanthroline) concept [53]. This approach is based on the kinetic control of the metal complexation equilibrium and, in the target complex, the Cu(I) ion turns out to be bound to a simple and a bulky phenanthroline ligand; this concept is schematically illustrated in Fig. 10.

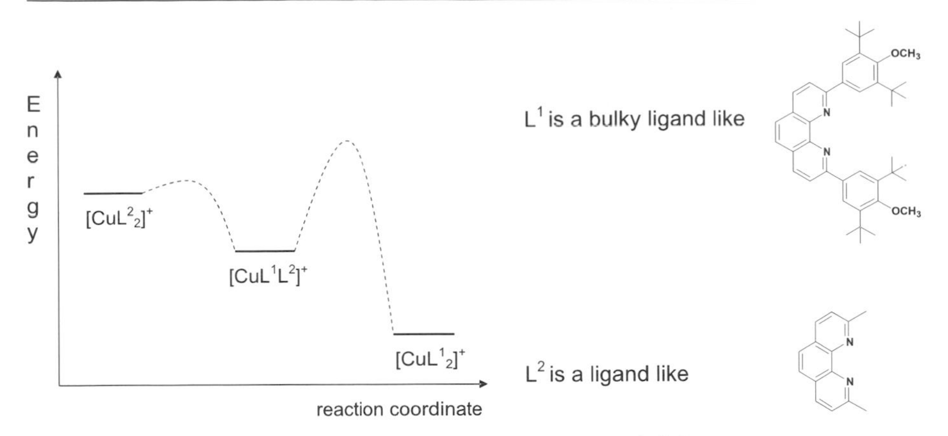

Fig. 10 Kinetic control in the formation of heteroleptic $[CuL^1L^2]^+$ complexes (HETPHEN approach)

Simple and supramolecular Cu(I)-bisphenanthroline complexes exhibit very interesting and largely tunable photophysical properties that will be illustrated in the next sections.

2 Cu(I)-Bisphenanthroline Complexes

2.1 Ground State Geometry

Cu(I)-bisphenanthroline complexes generally display distorted tetrahedral geometries. The distortion from D_{2d} symmetry can be visualized with the aid of Fig. 11 [54].

θ_x, θ_y, and θ_z define the interligand angles based on the CuN_4 core of the complex. When a molecule possesses a perfect tetrahedral geometry (D_{2d}), $\theta_x = \theta_y = \theta_z = 90°$, whereas the square planar geometry D_2 implies $\theta_x = \theta_y = 90°$ and $\theta_z = 0°$. Practically, θ_z is the dihedral angle between the ligand planes and a decrease from 90° indicates a flattening distortion of the molecule that progressively lowers its symmetry to D_2. The θ_x and θ_y values indicate the degree of "rocking" and "wagging" distortions [55].

These distortions are due to intra- and intermolecular (in solid state crystals) π - stacking interactions which also cause considerable displacement from D_{2d} symmetry. In practice, combinations of various types of distortions occur in $[Cu(NN)_2]^+$ complexes and their extent is dictated by the size, chemical nature, and positions of the phenanthroline substituents. Recently, the parameter ξ_{CD} has been proposed to quantify the degree of distortion as leading to a coordination geometry (and a spectrum) very similar to that of the much simpler $[Cu(\mathbf{1})_2]^+$ complex.

Some Cu(I)-bisphenanthroline compounds have also been utilized as receptors for dicarboxylic acids [76] and spherical inorganic anions [77]. The recognition motif is established by suitably functionalizing one phenyl residue in the 2 or 9 position of the phenanthroline chelating units, which are made able to pinch the pertinent substrate. The formation of the supramolecular adducts is monitored through the substantial changes of the MLCT absorption bands that occur as a consequence of the modification of the coordination geometry and related symmetry [76].

2.3 Excited State Distortion: Pulsed X-ray and Transient Absorption Spectroscopy

Upon light excitation of $[Cu(NN)_2]^+$ complexes the lowest MLCT excited state is populated, thus the metal center changes its formal oxidation state from Cu(I) to Cu(II) [15]. The Cu(I) MLCT excited complex undergoes further flattening compared to its ground state and assumes a geometry similar to that of ground state Cu(II)-bisphenanthroline complexes [59]. In this excited state flattened tetrahedral structure a fifth coordination site is made available for the Cu(II) d^9 ion, that can be filled by nucleophilic species such as solvent molecules and counterions. The intermediate species thus obtained is termed "pentacoordinated exciplex". The process, which is schematically depicted in Fig. 16, had been proposed by McMillin and coworkers nearly 20 years ago on the basis of classical photochemical experiments on series of $[Cu(NN)_2]^+$ complexes with increasingly nucleophilic counteranions [78].

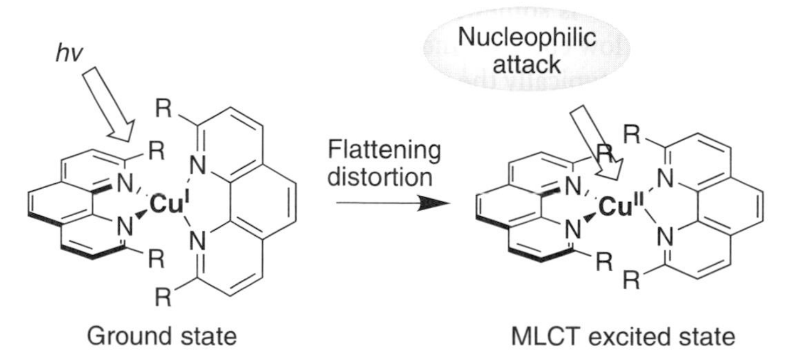

Fig. 16 Flattening distortion and subsequent nucleophilic attack by solvent, counterion, or other molecules following light excitation in Cu(I)-phenanthrolines. The size (and position) of the R substituents is of paramount importance in determining both the extent of the distortion and the protection of the newly formed Cu(II) ion from nucleophiles

In recent years this hypothesis has been nicely confirmed thanks to the development of light-initiated time-resolved X-ray absorption spectroscopy

(LITR-XAS) [79]. This pump-and-probe technique allows one to catch the transient oxidation state of a metal atom as well as its surrounding structures following photoexcitation via an ultrafast laser source; accordingly, it is particularly suited to investigating the process depicted in Fig. 16. Practically, the information obtained via LITR-XAS is a sort of snapshot of electronic excited states in disordered media (e.g. solution), which are generated via a UV-VIS femtosecond pump laser and subsequently probed with a 30–100 ps intense X-ray pulse produced by 3rd generation large synchrotron facilities [80].

This technique has been applied in solution studies to $[Cu(\mathbf{1})_2]^+$ and unambiguously confirmed that, in the thermally equilibrated MLCT excited state, the copper ion is pentacoordinated both in poorly donor (toluene) [81] or highly donor (CH_3CN) [82] solvents; in addition, the copper ion has the same oxidation state as the corresponding ground state Cu(II) complex in both cases. Analogous investigations have been carried out also on Cu(I) complexes as solid crystals ("photocrystallography") [55, 83]. LITR-XAS studies of $[Cu(\mathbf{1})_2]^+$ in solution have been complemented by optical time-resolved spectroscopy, which evidenced spectroscopic features in the ps timescale, associated to excited state structural rearrangements, possibly flattening distortion [82].

We have also carried out femtosecond transient absorption studies on $[Cu(\mathbf{7})_2]^+$ and $[Cu(\mathbf{8})_2]^+$ in CH_2Cl_2 (Fig. 17) [84]. These complexes are characterized by alkyl- and more cumbersome phenyl-residues in the 2 and 9 position of the phenanthroline ligand, which imparts rather different photophysical properties (i.e. shape of UV-VIS absorption, luminescence spectra, excited state lifetime) [85]. Despite this diversity, femtosecond transient absorption spectra have revealed a dynamic process lasting 15 ps in both cases Fig. 18.

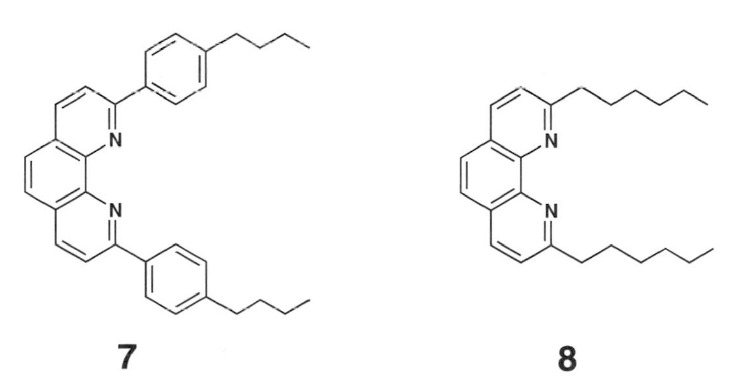

Fig. 17 Ligands **7** (2,9-bis(4-*n*-butylphenyl)-1,10-phenanthroline) and **8** (2,9-di-*n*-hexyl-1,10-phenanthroline)

Specific assignment of the observed spectral variation to (i) flattening distortion or (ii) extra ligand pick-up, two processes that might also occur sim-

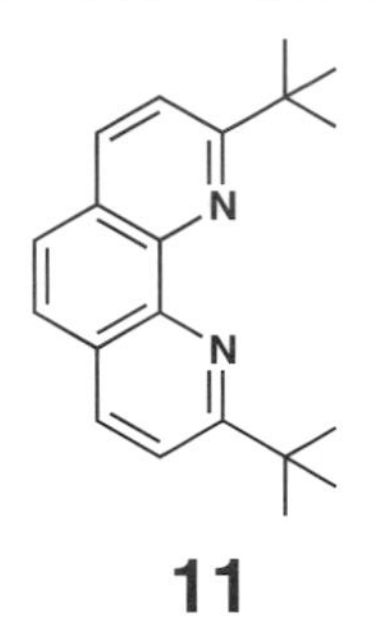

Fig. 20 Ligand 11 2,9-di-*tert*-butyl-1,10-phenanthroline

served large Stokes shift (over 5000 cm^{-1}) had attributed it to the lowest ^{3}MLCT excited state [91], likewise the popular family of octahedral Ru(II)-polypyridines [6]. McMillin and coworkers, instead, suggested that emission of $[Cu(NN)_2]^+$ compounds arises from two MLCT excited states in thermal equilibrium, i.e. a singlet (^{1}MLCT) and a triplet (^{3}MLCT) [92]. The energy gap between these states was found to be about 1500–2000 cm^{-1} and, at room temperature, the population of the lower lying ^{3}MLCT level exceeds that of ^{1}MLCT. At 77 K where the excited state population is largely frozen in the triplet, the emission band is red-shifted and much weaker compared to room temperature, a rather unusual trend. Recent studies have confirmed and refined this rationale [81, 85, 89].

A few years ago our group discussed detailed temperature-dependent luminescence studies of a series of $[Cu(NN)_2]^+$ complexes of 2,9-disubstituted phenanthroline ligands [85]. The above-described two-level model, which implies red-shift and intensity decrease of the emission band upon temperature lowering, is always obeyed except when long alkyl chains are utilized as substituents of the phenanthroline chelating agent. In this case the "regular" trend is obeyed only until the matrix remains fluid ($T > 150$ K) but, when the matrix becomes rigid ($T < 120$ K), a substantial blue shift and intensity increase is observed. At 95 K these compounds are bright emitters as intense as Ru(II) complexes [85]. The two trends (i.e. "classical" vs. "odd") in the luminescence intensity of $[Cu(NN)_2]^+$ as a function of temperature are illustrated in Fig. 21 for $[Cu(\mathbf{7})_2]^+$ and $[Cu(\mathbf{8})_2]^+$.

We rationalized the unusual behavior observed for complexes having ligands with long alkyl chains to steric, rather than electronic factors. This interpretation has been confirmed by Siddique et al. who found that the radiative constant (k_r) of the lowest ^{3}MLCT level is structure-sensitive and increases dramatically when the dihedral angle θ_z is approaching 90° [89]. In other words the long-alkyl chain blocks the ground state geometry in a rigid matrix and grants intense orange luminescence from the lowest ^{3}MLCT level. We recently observed that the 77 K intense luminescence can also be observed for homoleptic complexes of the phenyl-substituted ligand, such as **12**

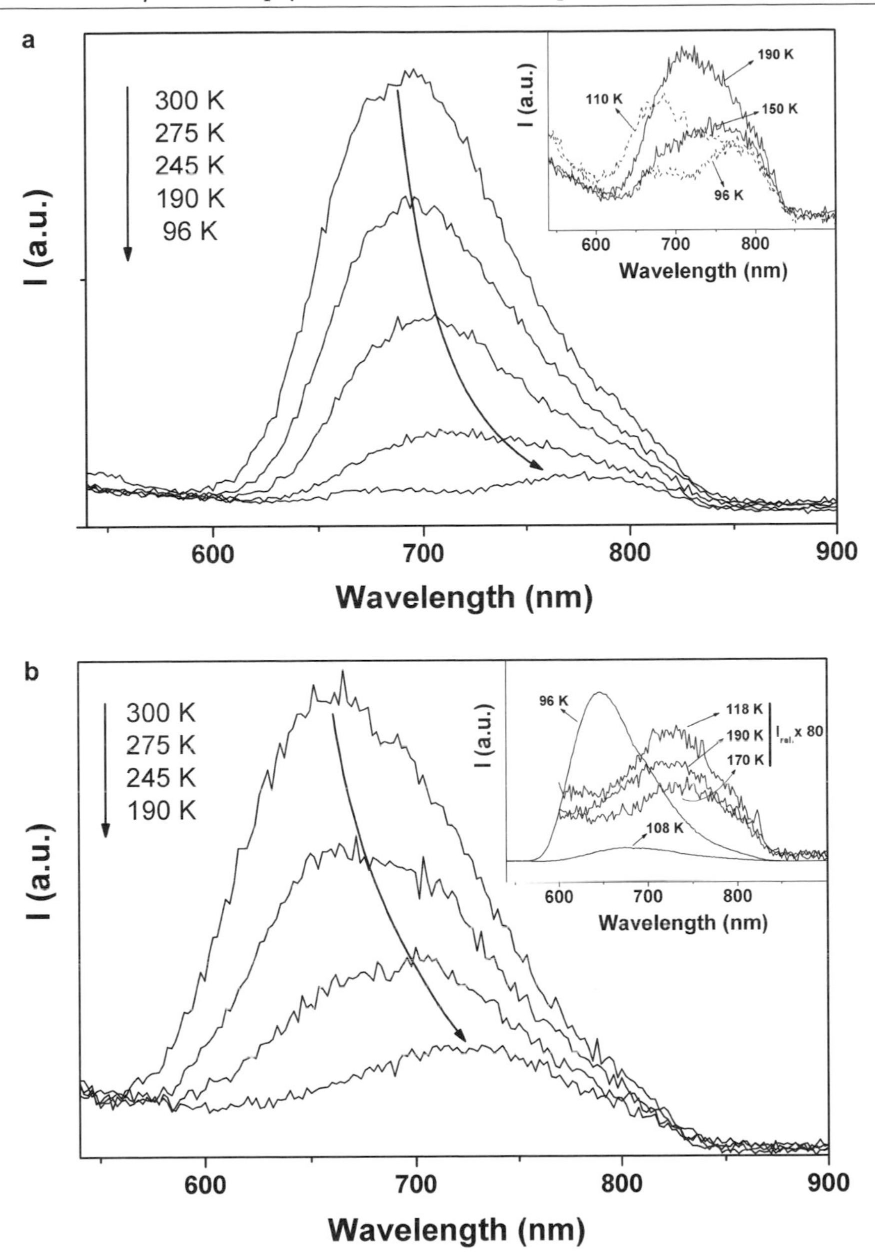

Fig. 21 Temperature dependence of the luminescence spectra of $[Cu(\mathbf{7})_2]^+$ (top) and $[Cu(\mathbf{8})_2]^+$ (bottom) in CH_2Cl_2 MeOH 1:1 (v/v). In the fluid domain (up to 170 K) emission intensity decrease and spectral red-shifting is observed by lowering temperature in both cases. By contrast when the solvent matrix becomes rigid (around 120 K), the two compounds behave differently. For the 2,9-dialkylphenanthroline complex (bottom panel) a complete reversal of the previous trend is observed with intensity recovery and blue shift. At 96 K a very strong luminescence band is recorded

tionalized derivative. The combined effect of these two factors (kinetic and thermodynamic) can explain the different and unexpected trend in photoprocesses of multicomponent arrays containing Cu(I)-phenanthrolines linked to methanofullerenes vs. bismethanofullerenes, which has been found in a variety of molecular architectures such as dendrimers [38], rotaxanes [98] and sandwich-type dyads [110].

Exhaustive review articles presenting photophysical investigations on fullerene- and porphyrin-type arrays built-up around $[Cu(NN)_2]^+$ centers have been published recently and we suggest the reader refers to these papers for a comprehensive and updated overview on this topic [15, 25, 111, 112].

2.6 Bimolecular Quenching Processes

Excited state electrochemical potentials can be obtained from the *ground state* monoelectronic electrochemical potentials and the spectroscopic energy (E^{oo} in eV units, to be considered divided by a unitary charge) related to the involved transition, according to Eqs 2 and 3 [6]:

$$E(A^+/{}^*A) = E(A^+/A) - E^{oo} \quad (2)$$

$$E({}^*A/A^-) = E(A/A^-) + E^{oo} \quad (3)$$

Hence the variation of the electron-donating or accepting capability of a given molecule A, upon light excitation, can be easily assessed. In Eqs 2 and 3: ${}^*A$ denotes the lowest-lying electronically excited state of A and its spectroscopic energy (E^{oo}) can be estimated from the onset of emission spectra [6].

Oxidation from Cu(I) to Cu(II) is easily accomplished and the MLCT excited states of Cu(I)-bisphenanthrolines are, therefore, potent reductants. For example $[Cu(\mathbf{3})_2]^+$ is a more powerful reductant than the very popular photosensitizer $[Ru(bpy)_3]^{2+}$ (A^+/A = – 1.11 and – 0.85 V, respectively) owing to its more favorable ground state 2+/+ potential (+ 0.69 vs. + 1.27 V), that largely compensates the lower content of excited state energy (1.80 vs. 2.12 eV) [15]. By contrast reduction of Cu(I)-bisphenanthrolines is strongly disfavored and they are mild excited state oxidants; accordingly, only a few examples of reductive quenching of $[Cu(NN)_2]^+$ complexes are reported in the literature, with ferrocenes as donors [113, 114].

Oxidative quenching of $[Cu(NN)_2]^+$'s by Co(III) and Cr(III) complexes as well as nitroaromatic compounds and viologens has been reported and comprehensively reviewed [115]. Some attempts to sensitize wide band-gap semiconductors with Cu(I) complexes were also carried out [115] but so far they do not seem to be competitive in terms of stability and efficiency with those based on Ru(II) complexes [12]. Energy transfer quenching to molecules possessing low-lying triplets such as anthracene has been demonstrated via transient absorption spectroscopy [116, 117], whereas oxygen quenching, which in principle can occur both via energy- and electron transfer, was evidenced by monitoring sensitized singlet oxygen luminescence in the NIR region [29, 36]. Optically pure dicopper trefoil knots with $[Cu(NN)_2]^+$-type cores have been reported to quench the emission of the Λ or Δ forms of Tb(III) and Eu(III) complexes, a very rare example of enantioselective luminescence quenching [118].

$[Cu(NN)_2]^+$ complexes have also been used as substrates for DNA binding, trying to take advantage of the sensitivity of the luminescence of Cu(I)-phenanthrolines to the local environment [63]. The structure of the associates has not been clarified: both electrostatic binding and intercalation of the aromatic ligands between adjacent bases are possible. Cu(I)-porphyrins seem to be more promising substrates for DNA [63].

3 Heteroleptic Diimine/Diphosphine $[Cu(NN)(PP)]^+$ Complexes

3.1 Photophysical Properties

Heteroleptic Cu(I) complexes containing both N- and P-coordinating ligands, $[Cu(NN)(PP)]^+$, have been studied since the late 1970s [119]. The replacement of one N-N ligand with a P-P unit is often aimed at improving the emission properties. Accordingly, the relentless quest for highly performing luminescent metal complexes [7] has sparked revived interest in these compounds in recent years [120–122].

The absorption and luminescence spectrum of $[Cu(dbp)(POP)]^+$ (dbp = 2,9-butyl-1,10-phenanthroline and POP = bis[2-(diphenylphosphino)phenyl] ether) is reported in Fig. 24, as a representative example for this class of compounds [123]. Substantial blue-shifts of the lower-energy bands are observed compared to typical spectra of $[Cu(NN)_2]^+$ compounds (see Sect. 2).

UV spectral features above 350 nm are due to ligand-centered transitions whereas those in the 350–450 nm window are attributed to MLCT levels. $[Cu(NN)(PP)]^+$ complexes are subject to dramatic oxygen quenching, as deduced from the strong difference in excited state lifetimes passing from air-equilibrated to oxygen-free CH_2Cl_2 solution, 250 ns and 17 600 ns in the case of $[Cu(dbp)(POP)]^+$ [123]. The character of the emitting state in $[Cu(NN)(PP)]^+$ complexes has been discussed since their first characterization [119] and now its MLCT nature is established experimentally and theoretically [120, 124, 125]. The electron-withdrawing effect of the P–P unit on the metal center tends to disfavor the Cu(I)→N–N electron donation, as also reflected by the higher oxidation potential of the Cu(I) center compared to $[Cu(NN)_2]^+$ compounds [126], leading to a blue shift of MLCT transitions. This, according to the energy gap law [127], explains the emission enhance-

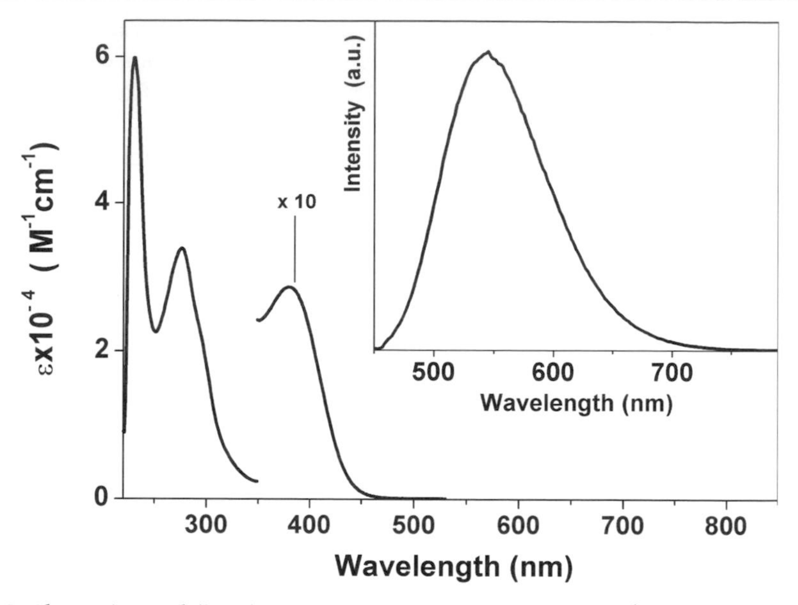

Fig. 24 Absorption and (inset) emission spectra of $[Cu(dbp)(POP)]^+$ in CH_2Cl_2

ment of $[Cu(NN)(PP)]^+$, that typically falls in the green spectral window, compared to weaker red-emitting $[Cu(NN)_2]^+$ complexes.

The luminescence efficiency of MLCT excited states in $[Cu(NN)(PP)]^+$ compounds is strongly solvent- and oxygen-dependent because it can be decreased by exciplex quenching [128, 129], in line with what is observed for the $[Cu(NN)_2]^+$ analogues (see above). Therefore, the geometry of $[Cu(NN)(PP)]^+$ complexes plays a central role in addressing the extent of luminescence efficiency, even though this is hard to predict a priori.

A variety of bidentate phosphine ligands has been prepared to coordinate Cu(I) in tandem with phenanthroline-type units: bis[2-(diphenylphosphino) phenyl]ether (POP), triphenylphosphine (PPh_3), bis(diphenylphosphino) ethane (dppe), and bis(diphenyl-phosphino)methane (dppm), represent some recent examples [120, 122, 123, 130, 131], Fig. 25.

Among them, the family of mononuclear $[Cu(phen)(POP)]^+$ complexes proposed by McMillin (see Fig. 25 for the PP-type ligands), where phen indicates a variably substituted 1,10 phenanthroline, shows an impressive emission efficiency compared to $[Cu(NN)_2]^+$ compounds [120]. Especially on passing from pristine phenanthroline to dimethyl- or diphenyl-substituted analogues, and thanks to the efficient steric and electron-withdrawing effects of the POP ligand, remarkable emission quantum yields ($\Phi_{em} \sim 0.15$ in CH_2Cl_2 oxygen-free solution) and long lifetimes ($\sim 15\,\mu s$) have been measured. On the contrary, the replacement of the POP ligand with two PPh_3 units, gives less remarkable results due to the lower geometric rigidity which leads to weak and red-shifted emissions comparable to those of

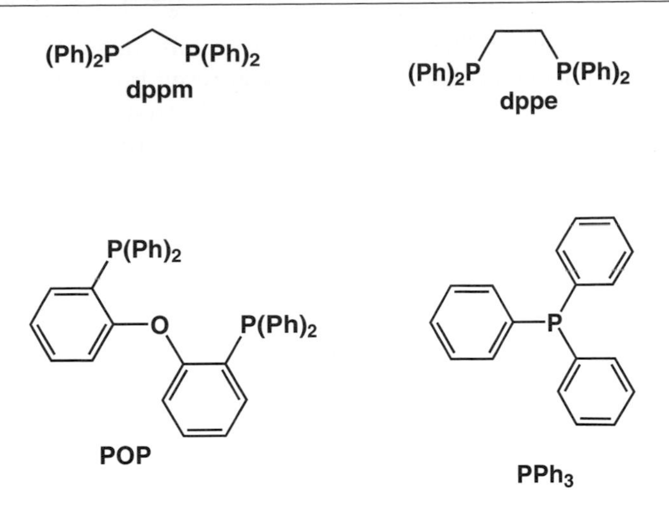

Fig. 25 Some ligands typically used as P–P units in $[Cu(NN)(PP)]^+$ complexes

bis-phenanthroline-type complexes. Importantly, the P–Cu–P angle decreases from 122.7° in $[Cu(dmp)(PPh_3)_2]^+$ to 116.4° in $[Cu(dmp)(POP)]^+$ also allowing easier access for exciplex quenching over the fifth coordination position in the former. This example highlights the importance of having both conditions (i.e. steric protection and increased electron-withdrawing character of the P–P ligand) simultaneously satisfied for optimized photoluminescence performance of $[Cu(NN)(PP)]^+$ compounds.

The importance of the choice of the P–P ligand for the coordination of the metal ion is evidenced also by the systems recently investigated by Wang et al. [121], in which ligands other than phenanthroline have been utilized (Fig. 26).

By keeping the N–N ligand unchanged, the luminescence properties of the complexes (solid matrix, RT) increase on passing from dppe to POP to, sur-

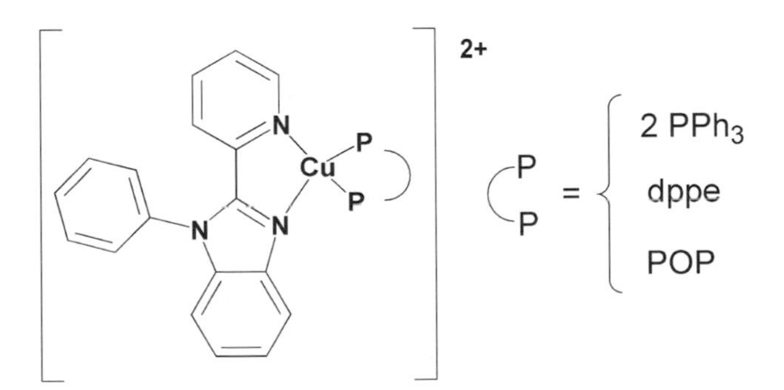

Fig. 26 General structure of $[Cu(ppb)(P)_2]$ complexes (pbb = 2-(2′-pyridyl)benzimidazolylbenzene)

prisingly, PPh_3. Although POP provides the best emission performance when combined with phenanthroline ligands, a better result is found here for PPh_3. This shows that subtle and combined steric and electronic effects of *both* P–P and N–N ligands are crucial for an enhanced light output, highlighting that general rules able to predict the photophysical behavior of $[Cu(NN)(PP)]^+$ complexes are not easy to draw. From an electronic point of view, both POP and PPh_3 units promote the usual blue shift of the MLCT state compared to $[Cu(NN)_2]^+$ compounds, as predictable by the substantially higher oxidation potential of the Cu(I) center of heteroleptic $[Cu(NN)(PP)]^+$ complexes.

Dinuclear Cu(I) complexes have also been synthesized and investigated, two of them (**A** and **B**) are depicted in Fig. 27. Despite the presence of a P–P-type ligand, complex **A** shows a luminescence band peaked at 700 nm with a lifetime of 320 ns in the solid state [132]. The X-ray crystal structure indicates a distorted tetrahedral geometry which, combined to the scarcely protective 2,5-bppz N–N ligand (2,5-bis(2-pyridil)pyrazine) leads to a weakly red-emitting compound. By changing the N-N ligand (Fig. 27, **B**), a stronger

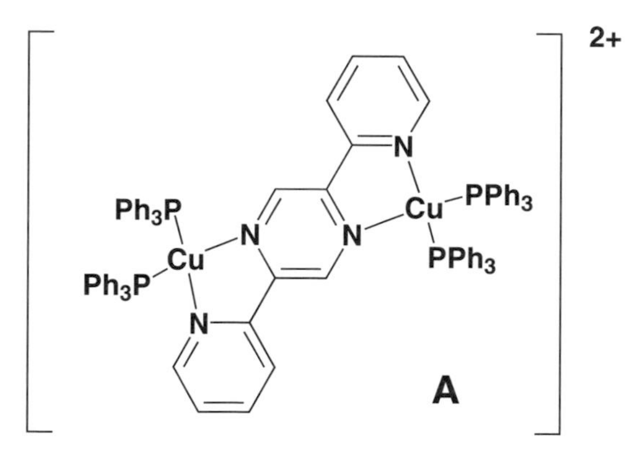

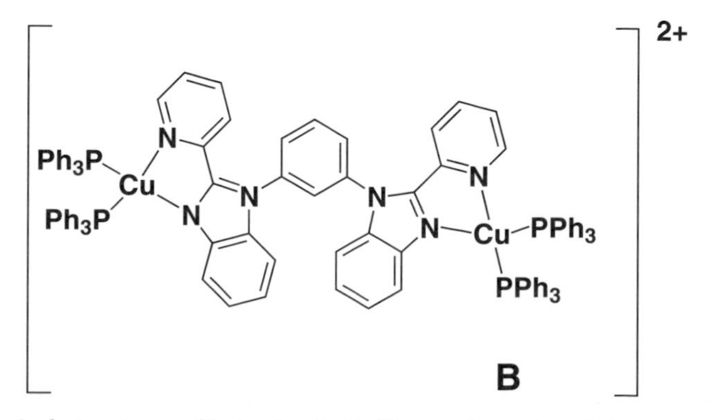

Fig. 27 Chemical structures of heteroleptic Cu(I) complexes **A** and **B**

green emission (λ_{max} = 550 nm) is detected with a quantum yield of 0.17 at 77 K in CH_2Cl_2 matrix [133]. For complex **B**, although the geometric structure is similar to **A**, the oxidation potentials of the N–N ligand are substantially greater, pushing the MLCT levels at higher energy.

3.2
OLED and LEC Devices

The outstanding photophysical performances of some $[Cu(NN)(PP)]^+$ complexes make them potentially attractive for optoelectronic devices requiring highly luminescent materials [134–137]. This interest is also related to the lower cost and higher relative abundance of copper compared to more classical emitting metals such as europium or iridium. Some research groups have recently fabricated OLED devices with $[Cu(NN)(PP)]^+$ complexes [138]. It has been shown that they can be profitably used as electrophosphorescent emitters and provide device efficiency comparable to that of Ir(III) complexes in similar device structures (11.0 cd/A at 1.0 mA/cm^2, 23% wt Cu(I)-complex dispersed in PVK matrix). Also Li et al. have obtained a highly efficient electrophosphorescent OLED with the complex reported in Fig. 28 [139].

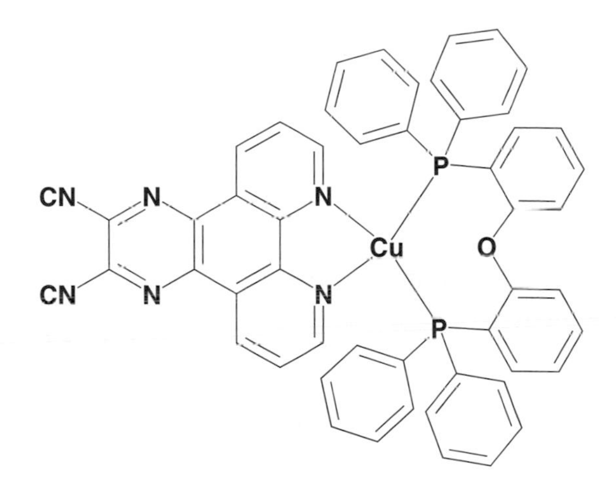

Fig. 28 The complex $[Cu(Dicnq)(POP)]^+BF_4$ used by Li et al. to make a highly efficient electroluminescent OLED device (Dicnq = 6,7-Dicyanodipyrido[2,2-*d*:2′,3′-*f*] quinoxaline)

The performances of OLEDs fabricated by the vacuum vapor deposition technique with this complex are among the best reported for devices incorporating Cu(I) complexes as emitters. A low turn-on voltage of 4 V, a maximum current efficiency up to 11.3 cd/A, and a peak brightness of 2322 cd/m^2 have been achieved.

A different type of electroluminescent device is a light-emitting electrochemical cell (LEC). LECs are substantially different from OLEDs due to the fact that mobile ions in the electroluminescent layer drift towards the electrodes when a voltage is applied over the device, thereby facilitating charge-carrier injection from the electrodes. This results in two important advantages compared to traditional OLEDs: (i) thick electroactive layers can be used without severe voltage penalties and shorts can be eliminated even for large-area pixels; (ii) matching of the work function of the electrodes with the energy levels of the electroluminescent material is not required.

We have recently described novel Cu(I) complexes with excellent PL performance (Q.Y. up to 0.28 in oxygen-free CH_2Cl_2) and the first LEC device made with a Cu(I) complex, Fig. 29 [123].

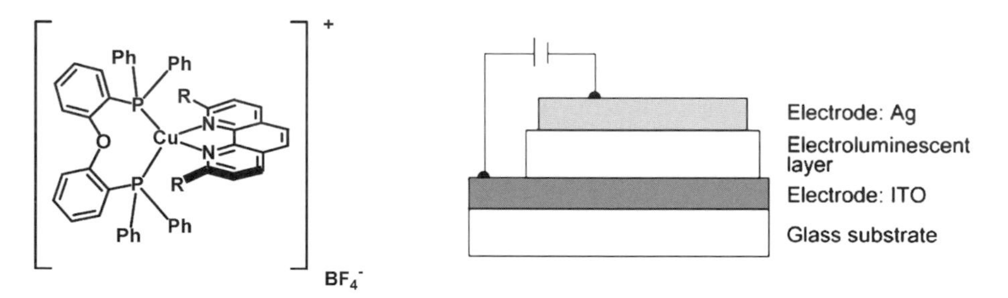

Fig. 29 Chemical structure of the complex used to make the first LEC device based on a Cu(I)-complex, R = *n*-butyl. A schematic representation of the device structure is also depicted; in the electroluminescent layer the complex is dispersed in a polymethylmetacrylate matrix (PMMA)

The device efficiency turned out to be moderate but comparable to LEC devices made with Ru(II)-type compounds [134]. Wang et al. used the same complex but, changing experimental conditions, could make a more efficient green light emitting device (CIE coordinates: 0.25, 0.60) with a maximum current efficiency of 56 cd/A at 4.0 V, corresponding to an external quantum yield of 16% [140]. This work notes the importance of the optimization of LEC device parameters such as the response time, which greatly depends on the counterion, driving voltage, and thickness of the emitting layer. Further efforts are needed to substantially improve the device stability and light output in order to take advantage of the low-cost and limited environmental damaging effects of copper materials.

Finally, it is worth pointing out that also the family of cuprous cluster (described in the next paragraph) has been tested in devices. In the late 1990s Ma et al. described the electroluminescence properties of a LED containing a tetranuclear Cu(I) cluster as the active component contributing to broaden the pool of electroluminescent materials outside the traditional boundaries of organic dyes and polymers [141].

4 Cuprous Clusters

4.1 Cuprous Halide Clusters

Cluster compounds contain a group of two or more metal atoms where direct and substantial metal-metal bonding is present. Cuprous halide clusters have been known for about 100 years [142], their general formula is $Cu_nX_nL_m$(X = Cl^-, Br^- or I^-; L = N or P belonging to an organic molecule). For instance, in solution, mixtures of Cu(I) salts, iodine (I) and pyridine-type molecules (py) are primarily present as tetrahedral clusters $Cu_4I_4py_4$ and give origin to mononuclear or dinuclear structures only if forced by mass action law under high pyridine concentration. Normally, copper(I) complexes in solution are quite labile towards ligand substitution and the formation of new species is driven by thermodynamic stability rather than kinetic control.

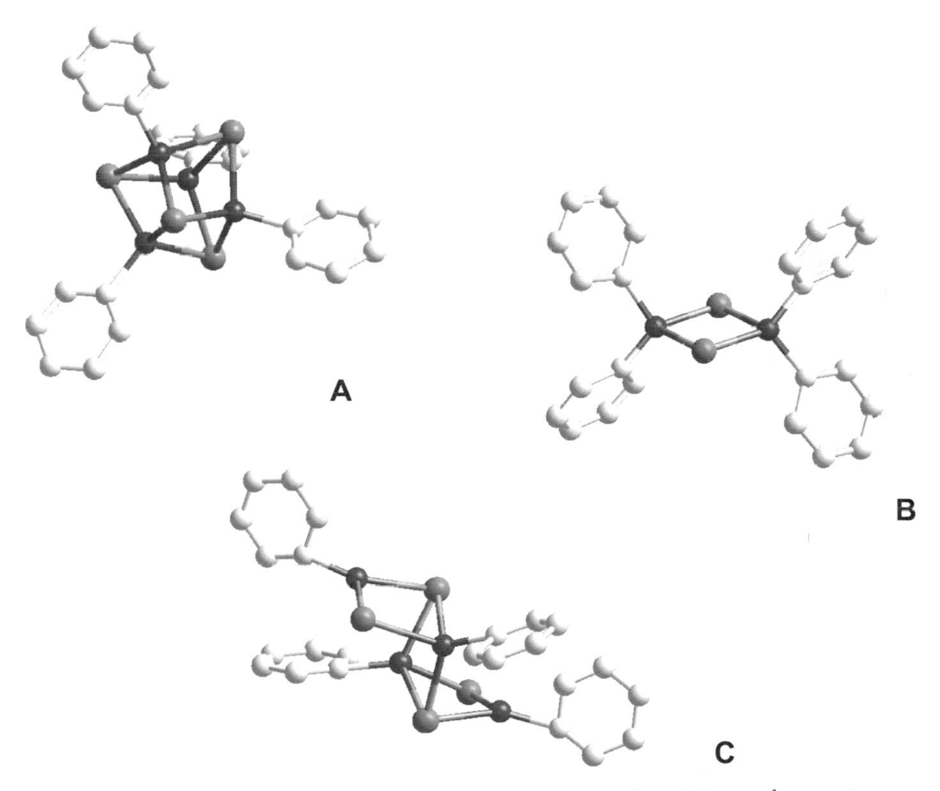

Fig. 30 Illustrations of $Cu_4I_4py_4$ (A), $Cu_2I_2py_4$ (B), $[Cu_3(\mu\text{-dppm})_3(\mu_3 - \eta^1\text{-C}\Xi\text{C-benzo-15-crown-5})_2]^+$ (C), and the repeating unit of a "stairstep" polymer $[CuIpy]_n$ (D) redrawn from the structural data. *Black circles* = copper atoms; *grey circles* = iodine; *white rings* = pyridine residues

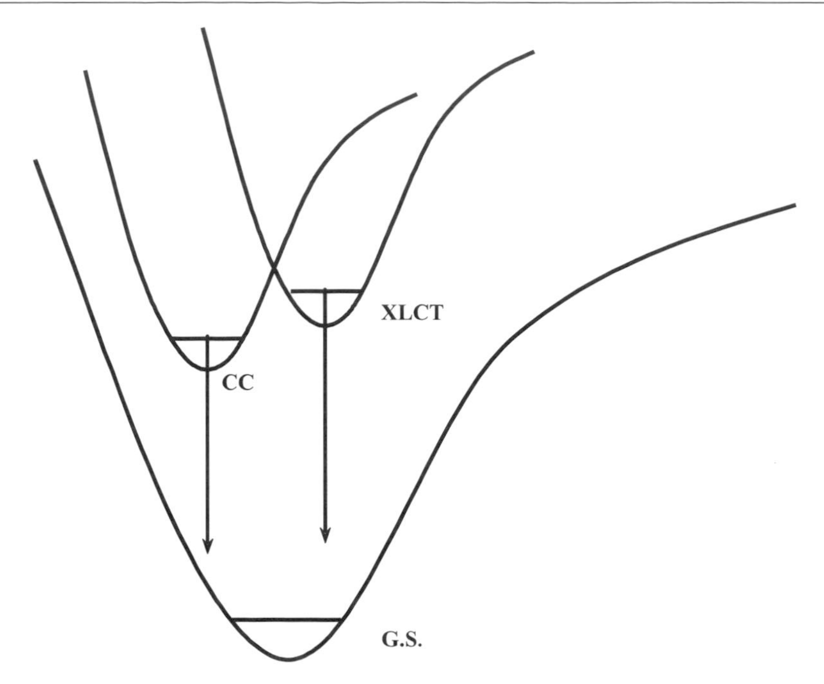

Fig. 32 Schematic representation describing the relative positions of the potential energy curves related to the emitting states of $Cu_4I_4py_4$

same conditions) also indicate that CC excited states are quite distorted relative to both the GS and XLCT levels. Such distortion was proposed as the cause for the lack of communication between CC and XLCT states, for the rigidochromism of the low CC energy band, and for the large reorganization energies of electron transfer reactions involving CC excited states [143].

In an interesting recent work on clusters of general formula CuXL (L = N–heteroaromatic ligands), it was shown that the energy of the emitting level can be finely tuned [149]. The emission is attributed to a MLCT charge transfer state, because no clear correlation between Cu–Cu distances and emission maxima was observed and also because the effects of bridging halides were smaller than those of *N*-heteroaromatic ligands, therefore the position of the luminescence band can be varied by increasing the electron-accepting character of the ligand L, Fig. 33. In these compounds Cu–Cu distances are in the range 2.9–3.3 Å, accordingly the weak metal interaction prevent cluster-centered luminescence. Very recently, density functional theory calculations have confirmed the involvement of the triplet cluster-centered (CC) and triplet XLCT excited states as the origin of the dual emission [151].

It must be pointed out, however, that short Cu–Cu separation does not automatically imply the establishment of metal-metal bonds and the effect of the bridging ligands has to be taken into account. For example Cotton et al.

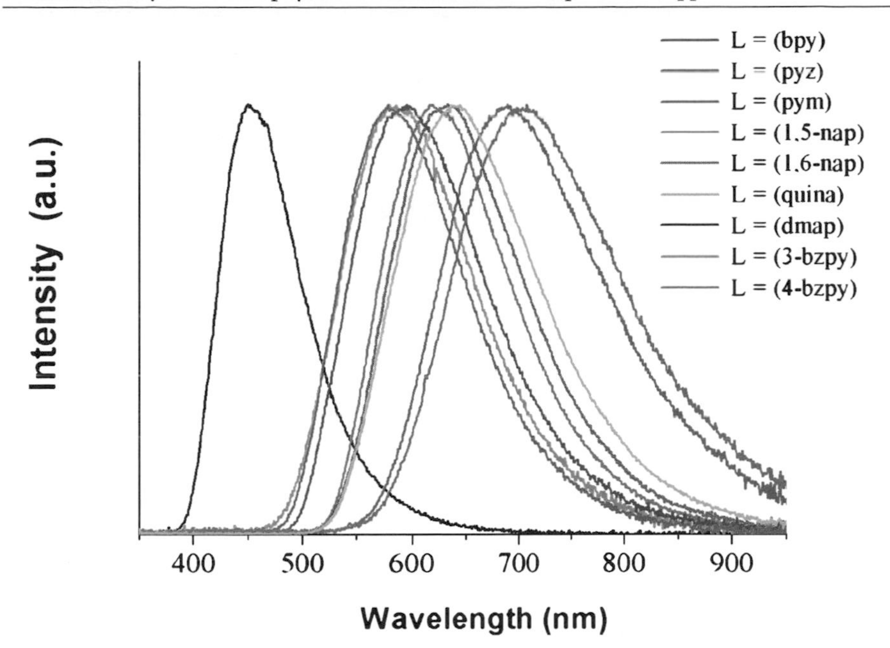

Fig. 33 Emission spectra of clusters CuXL (X = Br^-) at room temperature. The emission maxima of the complexes cover a wide range, from 450 to 740 nm depending on the *N*-heteroaromatic ligands. bpy = 4,4-bipyridine; pyz = pyrazine; pym = pymidine; pip = piperazine; 1,5-nap = 1,5-naphthyridine; 1,6-nap = 1,6-naphthyridine; quina = quinazoline; dmap = *N,N*-dimethyl-amino-pyridine; 3-bzpy = 3-benzoyl-pyridine; 4-bzpy = 4-benzoyl-pyridine. (Reprinted from [149] with permission, © (2006) American Chemical Society)

have carried out DTF calculations on $[Cu_2(hpp)_2]$, (where hpp^- = 1,3,4,6,7,8-hexahydro-2*H*pyrimido[1,2-*a*]pyrimidinate) to investigate the possibility of metal-metal bonding in a complex where short metal–metal separations are present [*d*Cu···Cu = 2.497(2) Å]. They concluded that there is no Cu–Cu bond and the short intermetal distance is related to the strong Cu–N bonds and the small bite angle of the bridging ligand [150].

4.3 Other Copper Clusters

Recently, the synthesis of several polynuclear copper(I) alkynyl clusters has been reported and their luminescence properties investigated in detail [152, 153]; these compounds exhibit intense and long-lived luminescence upon photoexcitation. For instance, the tetranuclear copper(I) alkynyl complex $[Cu_4(PPh_3)_4(L)_3]PF_6$, in Fig. 34 is characterized by an unusual open-cube structure, and exhibits a strong structured emission with two different max-

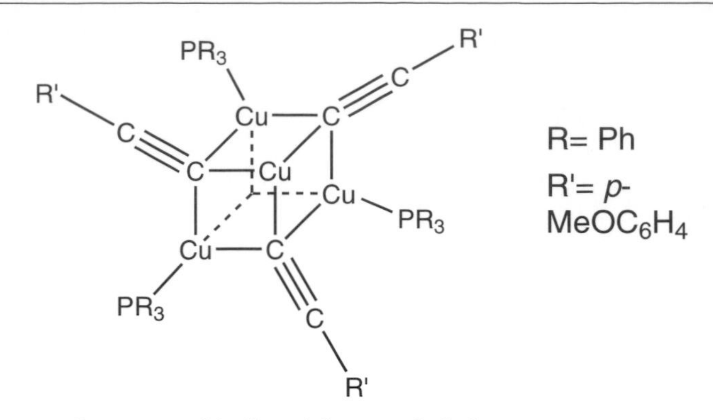

Fig. 34 A tetranuclear copper(I) alkynyl "open-cube" cluster

ima at 445 and 630 nm in solid state at 298 K and a single band with λ_{max} = 445 nm in rigid matrix at 77 K [154, 155].

For some of these open-cube compounds, an additional low-energy emission band at $\lambda > 623$–665 nm was observed in the solid-state spectra, similarly to what was observed for $[Cu_4I_4L_4]$ systems described above. In dichloromethane solution at ambient temperature they exhibit only an orange phosphorescence and the spectrum of $[Cu_4(PPh_3)_4(L)_3]PF_6$ (L = $p - n$OctC_6H_4) is depicted in Fig. 35 as a representative example for this class of compounds.

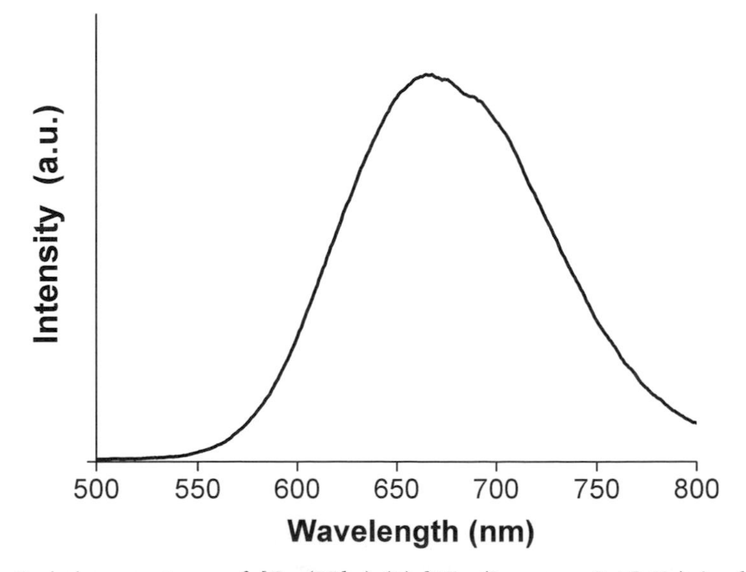

Fig. 35 Emission spectrum of $[Cu_4(PPh_3)_4(L)_3]PF_6$ (L = $p - n$OctC_6H_4) in degassed dichloromethane at 298 K (Reprinted from [155] with permission, © (2006) Wiley)

There are other examples of luminescent clusters, for instance trinuclear copper(I) pyrazolates displaying emission bands over a wide spectral range [156], or others with a core made of four Cu(I) and four sulfur atoms [157, 158]. There are examples of Cu(I) luminescent clusters with a higher nuclearity, some of them are heterometallic (6–8 metal centers) [159] while others are homometallic but with a higher nuclearity (16–20 metal centers). In the latter case, which bear alkynyl ligands [160], an emission band is observed in the UV spectral region. Finally, the possible use of copper cluster units to assemble polymeric compounds with a wide range of possible structures, from one- to three-dimensional should be noted [161, 162].

5 Miscellanea of Cu(I) Luminescent Complexes

In the previous sections we have presented the three main classes of Cu(I) compounds exhibiting interesting photophysical properties, namely Cu(I)-bisphenanthrolines, $[Cu(NN)(PP)]^+$ complexes and cuprous clusters. However, especially in recent years, a growing number of Cu(I) luminescent complexes with less conventional ligands have appeared in the literature and some of them will be now briefly presented.

The homoleptic Cu(I) complexes of the benzo[*h*]quinoline ligands (BHQ) depicted in Fig. 36 exhibit excellent luminescence properties in CH_2Cl_2 with quantum yields as high as 0.10 and $\tau = 5.3\ \mu s$ (ligand **C** in Fig. 36),

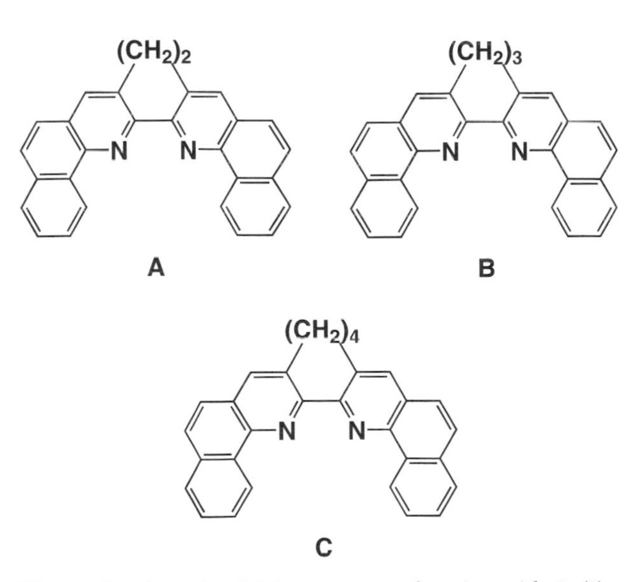

Fig. 36 Benzo[*h*]quinoline ligands which, upon complexation with Cu(I), provide highly luminescent complexes

Acknowledgements We thank the CNR (Progetto "Sistemi nanoorganizzati con proprietà elettroniche, fotoniche e magnetiche, commessa PM.P04.010 (MACOL)") and the EC through the Integrated Project OLLA (contract no. IST-2002-004607) for financial support. Over the years we worked on several collaborative projects related to Cu(I) complexes and, in this regard, we wish to thank Jean-François Nierengarten (Toulouse, France), Jean-Pierre Sauvage (Strasbourg, France) and Michael Schmittel (Siegen, Germany) along with many other colleagues from their research groups, whose names are cited in the references.

References

1. US Geological Survey (2006) Mineral Commodity Summaries http://minerals.er.usgs.gov/minerals/pubs/commodity/copper/, last visited: January 2006
2. Gordon RB, Bertram M, Graedel TE (2006) Proc Natl Acad Sci USA 103:1209–1214
3. Horvath O (1994) Coord Chem Rev 135:303–324
4. Sykora J (1997) Coord Chem Rev 159:95–108
5. Jørgensen CK (1963) Adv Chem Phys 5:33–146
6. Juris A, Balzani V, Barigelletti F, Campagna S, Belser P, von Zelewsky A (1988) Coord Chem Rev 84:85–277
7. Lowry MS, Bernhard S (2006) Chem Eur J 12:7970–7977
8. Balzani V, Juris A, Venturi M, Campagna S, Serroni S (1996) Chem Rev 96:759–833
9. Roundhill DM (1994) Photochemistry and Photophysics of Metal Complexes. Plenum Press, NY
10. Bignozzi CA, Argazzi R, Kleverlaan CJ (2000) Chem Soc Rev 29:87–96
11. Balzani V, Ceroni P, Juris A, Venturi M, Campagna S, Puntoriero F, Serroni S (2001) Coord Chem Rev 219:545–572
12. Grätzel M (2005) Inorg Chem 44:6841–6851
13. Vos JG, Kelly JM (2006) Dalton Trans pp 4869–4883
14. Kober EM, Caspar JV, Lumpkin RS, Meyer TJ (1986) J Phys Chem 90:3722–3734
15. Armaroli N (2001) Chem Soc Rev 30:113–124
16. Maestri M, Armaroli N, Balzani V, Constable EC, Thompson A (1995) Inorg Chem 34:2759–2767
17. Lippard SJ, Berg JM (1994) Principles of Bioinorganic Chemistry. University Science Books, Mill Valley, California
18. Holm RH, Kennepohl P, Solomon EI (1996) Chem Rev 96:2239–2314
19. Colman PM, Freeman HC, Guss JM, Murata M, Norris VA, Ramshaw JAM, Venkatappa MP (1978) Nature 272:319–324
20. Solomon EI (2006) Inorg Chem 45:8012–8025
21. Gewirth AA, Solomon EI (1988) J Am Chem Soc 110:3811–3819
22. Marcus RA, Sutin N (1985) Biochim Biophys Acta 811:265–322
23. Babcock GT, Wikstrom M (1992) Nature 356:301–309
24. Gamelin DR, Randall DW, Hay MT, Houser RP, Mulder TC, Canters GW, de Vries S, Tolman WB, Lu Y, Solomon EI (1998) J Am Chem Soc 120:5246–5263
25. Armaroli N (2003) Photochem Photobiol Sci 2:73–87
26. Schmittel M, Kalsani V (2005) Top Curr Chem 245:1–53
27. Sammes PG, Yahioglu G (1994) Chem Soc Rev 23:327–334
28. Cunningham CT, Moore JJ, Cunningham KLH, Fanwick PE, McMillin DR (2000) Inorg Chem 39:3638–3644

29. Armaroli N, Balzani V, Barigelletti F, De Cola L, Flamigni L, Sauvage JP, Hemmert C (1994) J Am Chem Soc 116:5211–5217
30. Dietrich-Buchecker C, Colasson B, Fujita M, Hori A, Geum N, Sakamoto S, Yamaguchi K, Sauvage JP (2003) J Am Chem Soc 125:5717–5725
31. Frey J, Kraus T, Heitz V, Sauvage JP (2005) Chem Commun, pp 5310–5312
32. Armaroli N, Balzani V, Collin JP, Gaviña P, Sauvage JP, Ventura B (1999) J Am Chem Soc 121:4397–4408
33. Weber N, Hamann C, Kern JM, Sauvage JP (2003) Inorg Chem 42:6780–6792
34. Baranoff E, Griffiths K, Collin JP, Sauvage JP, Ventura B, Flamigni L (2004) New J Chem 28:1091–1095
35. Kraus T, Budesinsky M, Cvacka JC, Sauvage JP (2006) Angew Chem Int Ed 45:258–261
36. Dietrich-Buchecker CO, Nierengarten JF, Sauvage JP, Armaroli N, Balzani V, De Cola L (1993) J Am Chem Soc 115:11237–11244
37. Perret-Aebi LE, von Zelewsky A, Dietrich-Buchecker CD, Sauvage JP (2004) Angew Chem Int Ed 43:4482–4485
38. Armaroli N, Boudon C, Felder D, Gisselbrecht JP, Gross M, Marconi G, Nicoud JF, Nierengarten JF, Vicinelli V (1999) Angew Chem Int Ed 38:3730–3733
39. Gumienna-Kontecka E, Rio Y, Bourgogne C, Elhabiri M, Louis R, Albrecht-Gary AM, Nierengarten JF (2004) Inorg Chem 43:3200–3209
40. Heuft MA, Fallis AG (2002) Angew Chem Int Ed 41:4520–4523
41. Cardinali F, Mamlouk H, Rio Y, Armaroli N, Nierengarten JF (2004) Chem Commun, pp 1582–1583
42. Zong RF, Thummel RP (2005) Inorg Chem 44:5984–5986
43. Ziessel R, Charbonniere L, Cesario M, Prange T, Nierengarten H (2002) Angew Chem Int Ed 41:975–979
44. Sauvage J-P, Dietrich-Buchecker CO (eds) (1999) Molecular Catenanes, Rotaxanes and Knots. A Journey through the World of Molecular Topology. Wiley VCH, Weinheim, Germany
45. Jimenez-Molero MC, Dietrich-Buchecker C, Sauvage JP (2002) Chem Eur J 8:1456–1466
46. Livoreil A, Sauvage JP, Armaroli N, Balzani V, Flamigni L, Ventura B (1997) J Am Chem Soc 119:12114–12124
47. Sauvage JP (2005) Chem Commun, pp 1507–1510
48. Bonnet S, Collin JP, Koizumi M, Mobian P, Sauvage JP (2006) Adv Mater 18:1239–1250
49. Kalsani V, Bodenstedt H, Fenske D, Schmittel M (2005) Eur J Inorg Chem 1841–1849
50. Schmittel M, Kalsani V, Fenske D, Wiegrefe A (2004) Chem Commun, pp 490–491
51. Schmittel M, Ammon H, Kalsani V, Wiegrefe A, Michel C (2002) Chem Commun, pp 2566–2567
52. Kalsani V, Ammon H, Jäckel F, Rabe JP, Schmittel M (2004) Chem-Eur J 10:5481–5492
53. Schmittel M, Ganz A (1997) Chem Commun, pp 999–1000
54. Dobson JF, Green BE, Healy PC, Kennard CHL, Pakawatchai C, White AH (1984) Aust J Chem 37:649–659
55. Coppens P, Vorontsov II, Graber T, Kovalevsky AY, Chen YS, Wu G, Gembicky M, Novozhilova IV (2004) J Am Chem Soc 126:5980–5981
56. Kovalevsky AY, Gembicky M, Coppens P (2004) Inorg Chem 43:8282–8289
57. Kovalevsky AY, Gembicky M, Novozhilova IV, Coppens P (2003) Inorg Chem 42:8794–8802

58. Miller MT, Gantzel PK, Karpishin TB (1998) Angew Chem Int Ed Engl 37:1556–1558
59. Miller MT, Gantzel PK, Karpishin TB (1998) Inorg Chem 37:2285–2290
60. Armaroli N, De Cola L, Balzani V, Sauvage JP, Dietrich-Buchecker CO, Kern JM (1992) J Chem Soc Faraday Trans 88:553–556
61. Zgierski MZ (2003) J Chem Phys 118:4045–4051
62. McMillin DR, Buckner MT, Ahn BT (1977) Inorg Chem 16:943–945
63. McMillin DR, McNett KM (1998) Chem Rev 98:1201–1219
64. Scaltrito DV, Thompson DW, O'Callaghan JA, Meyer GJ (2000) Coord Chem Rev 208:243–266
65. Federlin P, Kern JM, Rastegar A, Dietrich-Buchecker C, Marnot PA, Sauvage JP (1990) New J Chem 14:9–12
66. Gordon KC, McGarvey JJ (1991) Inorg Chem 30:2986–2989
67. Armaroli N, Rodgers MAJ, Ceroni P, Balzani V, Dietrich-Buchecker CO, Kern JM, Bailal A, Sauvage JP (1995) Chem Phys Lett 241:555–558
68. Ichinaga AK, Kirchhoff JR, McMillin DR, Dietrich-Buchecker CO, Marnot PA, Sauvage JP (1987) Inorg Chem 26:4290–4292
69. Phifer CC, McMillin DR (1986) Inorg Chem 25:1329–1333
70. Everly RM, McMillin DR (1991) J Phys Chem 95:9071–9075
71. Cunningham CT, Cunningham KLH, Michalec JF, McMillin DR (1999) Inorg Chem 38:4388–4392
72. Miller MT, Gantzel PK, Karpishin TB (1999) Inorg Chem 38:3414–3422
73. Miller MT, Karpishin TB (1999) Inorg Chem 38:5246–5249
74. Kalsani V, Schmittel M, Listorti A, Accorsi G, Armaroli N (2006) Inorg Chem 45:2061–2067
75. Gushurst AKI, McMillin DR, Dietrich-Buchecker CO, Sauvage JP (1989) Inorg Chem 28:4070–4072
76. Goodman MS, Hamilton AD, Weiss J (1995) J Am Chem Soc 117:8447–8455
77. Amendola V, Boiocchi M, Colasson B, Fabbrizzi L (2006) Inorg Chem 45:6138–6147
78. Everly RM, McMillin DR (1989) Photochem Photobiol 50:711–716
79. Chen LX (2005) Annu Rev Phys Chem 56:221–254
80. Chen LX (2004) Angew Chem Int Ed 43:2886–2905
81. Chen LX, Jennings G, Liu T, Gosztola DJ, Hessler JP, Scaltrito DV, Meyer GJ (2002) J Am Chem Soc 124:10861–10867
82. Chen LX, Shaw GB, Novozhilova I, Liu T, Jennings G, Attenkofer K, Meyer GJ, Coppens P (2003) J Am Chem Soc 125:7022–7034
83. Coppens P (2003) Chem Commun pp 1317–1320
84. Gunaratne T, Rodgers MAJ, Felder D, Nierengarten JF, Accorsi G, Armaroli N (2003) Chem Commun pp 3010–3011
85. Felder D, Nierengarten JF, Barigelletti F, Ventura B, Armaroli N (2001) J Am Chem Soc 123:6291–6299
86. Cody J, Dennisson J, Gilmore J, VanDerveer DG, Henary MM, Gabrielli A, Sherrill CD, Zhang YY, Pan CP, Burda C, Fahrni CJ (2003) Inorg Chem 42:4918–4929
87. Blaskie MW, McMillin DR (1980) Inorg Chem 19:3519–3522
88. Williams RM, De Cola L, Hartl F, Lagref JJ, Planeix JM, De Cian A, Hosseini MW (2002) Coord Chem Rev 230:253–261
89. Siddique ZA, Yamamoto Y, Ohno T, Nozaki K (2003) Inorg Chem 42:6366–6378
90. Miller MT, Gantzel PK, Karpishin TB (1999) J Am Chem Soc 121:4292–4293
91. Parker WL, Crosby GA (1989) J Phys Chem 93:5692–5696
92. Kirchhoff JR, Gamache RE, Blaskie MW, Del Paggio AA, Lengel RK, McMillin DR (1983) Inorg Chem 22:2380–2384

93. Cardenas DJ, Collin JP, Gaviña P, Sauvage JP, De Cian A, Fischer J, Armaroli N, Flamigni L, Vicinelli V, Balzani V (1999) J Am Chem Soc 121:5481–5488
94. Flamigni L, Talarico AM, Chambron JC, Heitz V, Linke M, Fujita N, Sauvage JP (2004) Chem-Eur J 10:2689–2699
95. Armaroli N, Balzani V, Barigelletti F, Decola L, Sauvage JP, Hemmert C (1991) J Am Chem Soc 113:4033–4035
96. Armaroli N, Balzani V, De Cola L, Hemmert C, Sauvage JP (1994) New J Chem 18:775–782
97. Dietrich-Buchecker CO, Sauvage JP, Armaroli N, Ceroni P, Balzani V (1996) Angew Chem Int Ed Engl 35:1119–1121
98. Armaroli N, Diederich F, Dietrich-Buchecker CO, Flamigni L, Marconi G, Nierengarten JF, Sauvage JP (1998) Chem-Eur J 4:406–416
99. Sandanayaka ASD, Watanabe N, Ikeshita KI, Araki Y, Kihara N, Furusho Y, Ito O, Takata T (2005) J Phys Chem B 109:2516–2525
100. Li K, Bracher PJ, Guldi DM, Herranz MA, Echegoyen L, Schuster DI (2004) J Am Chem Soc 126:9156–9157
101. Li K, Schuster DI, Guldi DM, Herranz MA, Echegoyen L (2004) J Am Chem Soc 126:3388–3389
102. Watanabe N, Kihara N, Furusho Y, Takata T, Araki Y, Ito O (2003) Angew Chem Int Ed 42:681–683
103. Linke M, Chambron SC, Heitz V, Sauvage SP, Encinas S, Barigelletti F, Flamigni L (2000) J Am Chem Soc 122:11834–11844
104. Andersson M, Linke M, Chambron JC, Davidsson J, Heitz V, Hammarström L, Sauvage JP (2002) J Am Chem Soc 124:4347–4362
105. Andersson M, Linke M, Chambron JC, Davidsson J, Heitz V, Sauvage JP, Hammarström L (2000) J Am Chem Soc 122:3526–3527
106. Flamigni L, Armaroli N, Barigelletti F, Chambron JC, Sauvage JP, Solladié N (1999) New J Chem 23:1151–1158
107. Chambron JC, Harriman A, Heitz V, Sauvage JP (1993) J Am Chem Soc 115:6109–6114
108. Chambron JC, Harriman A, Heitz V, Sauvage JP (1993) J Am Chem Soc 115:7419–7425
109. Holler M, Cardinali F, Mamlouk H, Nierengarten JF, Gisselbrecht JP, Gross M, Rio Y, Barigelletti F, Armaroli N (2006) Tetrahedron 62:2060–2073
110. Rio Y, Enderlin G, Bourgogne C, Nierengarten JF, Gisselbrecht JP, Gross M, Accorsi G, Armaroli N (2003) Inorg Chem 42:8783–8793
111. Clifford JN, Accorsi G, Cardinali F, Nierengarten JF, Armaroli N (2006) C R Chim 9:1005–1013
112. Flamigni L, Heitz V, Sauvage JP (2006) Struct Bond 121:217–261
113. Cunningham KL, McMillin DR (1998) Inorg Chem 37:4114–4119
114. Cunningham KL, Hecker CR, McMillin DR (1996) Inorg Chim Acta 242:143–147
115. Ruthkosky M, Kelly CA, Castellano FN, Meyer GJ (1998) Coord Chem Rev 171:309–322
116. Ruthkosky M, Castellano FN, Meyer GJ (1996) Inorg Chem 35:6406–6412
117. Castellano FN, Ruthkosky M, Meyer GJ (1995) Inorg Chem 34:3–4
118. Meskers SCJ, Dekkers H, Rapenne G, Sauvage JP (2000) Chem-Eur J 6:2129–2134
119. Buckner MT, McMillin DR (1978) J Chem Soc-Chem Commun, pp 759–761
120. Cuttell DG, Kuang SM, Fanwick PE, McMillin DR, Walton RA (2002) J Am Chem Soc 124:6–7
121. McCormick T, Jia WL, Wang SN (2006) Inorg Chem 45:147–155
122. Tsukuda T, Nakamura A, Arai T, Tsubomura T (2006) Bull Chem Soc Jpn 79:288–290

123. Armaroli N, Accorsi G, Holler M, Moudam O, Nierengarten JF, Zhou Z, Wegh RT, Welter R (2006) Adv Mater 18:1313–1316
124. Yang L, Feng JK, Ren AM, Zhang M, Ma YG, Liu XD (2005) Eur J Inorg Chem 1867–1879
125. Howell SL, Gordon KC (2004) J Phys Chem A 108:2536–2544
126. Kuang SM, Cuttell DG, McMillin DR, Fanwick PE, Walton RA (2002) Inorg Chem 41:3313–3322
127. Englman R, Jortner J (1970) Mol Phys 18:145–164
128. Rader RA, McMillin DR, Buckner MT, Matthews TG, Casadonte DJ, Lengel RK, Whittaker SB, Darmon LM, Lytle FE (1981) J Am Chem Soc 103:5906–5912
129. Palmer CEA, McMillin DR, Kirmaier C, Holten D (1987) Inorg Chem 26:3167–3170
130. Tsubomura T, Takahashi N, Saito K, Tsukuda T (2004) Chem Lett 33:678–679
131. Saito K, Arai T, Takahashi N, Tsukuda T, Tsubomura T (2006) Dalton Trans pp 4444–4448
132. Tsubomura T, Enoto S, Endo S, Tamane T, Matsumoto K, Tsukuda T (2005) Inorg Chem 44:6373–6378
133. Jia WL, McCormick T, Tao Y, Lu JP, Wang SN (2005) Inorg Chem 44:5706–5712
134. Slinker J, Bernards D, Houston PL, Abruna HD, Bernhard S, Malliaras GG (2003) Chem Commun, pp 2392–2399
135. Schubert EF, Kim JK (2005) Science 308:1274–1278
136. Bolink HJ, Cappelli L, Coronado E, Gavina P (2005) Inorg Chem 44:5966–5968
137. Holder E, Langeveld BMW, Schubert US (2005) Adv Mater 17:1109–1121
138. Zhang QS, Zhou QG, Cheng YX, Wang LX, Ma DG, Jing XB, Wang FS (2004) Adv Mater 16:432–436
139. Che GB, Su ZS, Li WL, Chu B, Li MT, Hu ZZ, Zhang ZQ (2006) Appl Phys Lett 89:103511
140. Zhang QS, Zhou QG, Cheng YX, Wang LX, Ma DG, Jing XB, Wang FS (2006) Adv Funct Mater 16:1203–1208
141. Ma YG, Che CM, Chao HY, Zhou XM, Chan WH, Shen JC (1999) Adv Mater 11:852–857
142. Raston CL, White AH (1976) J Chem Soc Dalton Trans 21:2153–2156
143. Vitale M, Ford PC (2001) Coord Chem Rev 219:3–16
144. Hardt HD, Pierre A (1973) Z Anorg Allg Chem 402:107
145. Ford PC, Cariati E, Bourassa J (1999) Chem Rev 99:3625–3647
146. Eitel E, Oelkrug D, Hiller W, Strahle J (1980) Z Naturforsch (B) 35:1247–1253
147. Kyle KR, Ryu CK, DiBenedetto JA, Ford PC (1991) J Am Chem Soc 113:2954–2965
148. Rath NP, Holt EM, Tanimura K (1986) J Chem Soc-Dalton Trans pp 2303–2310
149. Araki H, Tsuge K, Sasaki Y, Ishizaka S, Kitamura N (2005) Inorg Chem 44:9667–9675
150. Cotton FA, Feng XJ, Timmons DJ (1998) Inorg Chem 37:4066–4069
151. De Angelis F, Fantacci S, Sgamellotti A, Cariati E, Ugo R, Ford PC (2006) Inorg Chem 45:10576–10584
152. Yam VWW, Lo KKW, Wong KMC (1999) J Organomet Chem 578:3–30
153. Yam VWW (2002) Acc Chem Res 35:555–563
154. Yam VWW, Choi SWK, Chan CL, Cheung KK (1996) Chem Commun, pp 2067–2068
155. Chan CL, Cheung KK, Lam WH, Cheng ECC, Zhu N, Choi SWK, Yam VWW (2006) Chem-Asian J 1–2:273
156. Dias HVR, Diyabalanage HVK, Eldabaja MG, Elbjeirami O, Rawashdeh-Omary MA, Omary MA (2005) J Am Chem Soc 127:7489–7501
157. Che CM, Xia BH, Huang JS, Chan CK, Zhou ZY, Cheung KK (2001) Chem-Eur J 7:3998–4006

158. Kharenko OA, Kennedy DC, Demeler B, Maroney MJ, Ogawa MY (2005) J Am Chem Soc 127:7678–7679
159. Wei QH, Yin GQ, Zhang LY, Shi LX, Mao ZW, Chen ZN (2004) Inorg Chem 43:3484–3491
160. Baxter CW, Higgs AC, Jones AC, Parsons S, Bailey PJ, Tasker PA (2002) J Chem Soc Dalton Trans 4395–4401
161. Peng R, Li D, Wu T, Zhou XP, Ng SW (2006) Inorg Chem 45:4035–4046
162. He X, Lu CZ, Wu CD, Chen LJ (2006) Eur J Inorg Chem, pp 2491–2503
163. Riesgo EC, Hu YZ, Bouvier F, Thummel RP, Scaltrito DV, Meyer GJ (2001) Inorg Chem 40:3413–3422
164. Riesgo EC, Hu YZ, Thummel RP (2003) Inorg Chem 42:6648–6654
165. Zhang XM, Tong ML, Gong ML, Lee HK, Luo L, Li KF, Tong YX, Chen XM (2002) Chem-Eur J 8:3187–3194
166. Zheng SL, Zhang JP, Chen XM, Huang ZL, Lin ZY, Wong WT (2003) Chem-Eur J 9:3888–3896
167. Kunkely H, Vogler A (2003) Inorg Chem Commun 6:543–545
168. Pawlowski V, Knor G, Lennartz C, Vogler A (2005) Eur J Inorg Chem 3167–3171
169. Kinoshita I, Hamazawa A, Nishioka T, Adachi H, Suzuki H, Miyazaki Y, Tsuboyama A, Okada S, Hoshino M (2003) Chem Phys Lett 371:451–457
170. Song DT, Jia WL, Wu G, Wang SN (2005) Dalton Trans pp 433–438
171. Zhao SB, Wang RY, Wang SN (2006) Inorg Chem 45:5830–5840
172. Fournier E, Lebrun F, Drouin M, Decken A, Harvey PD (2004) Inorg Chem 43:3127–3135
173. Omary MA, Rawashdeh-Omary MA, Diyabalanage HVK, Rasika Dias HV (2003) Inorg Chem 42:8612–8614
174. Tsuboyama A, Okada S, Takiguchi T, Igawa S, Kamatani J, Furugori M, Canon KK (2005) JP Patent n. US Patent 2005014024

Top Curr Chem (2007) 280: 117–214
DOI 10.1007/128_2007_133

Published online: 27 June 2007

Photochemistry and Photophysics of Coordination Compounds: Ruthenium

Sebastiano Campagna[1] (✉) · Fausto Puntoriero[1] · Francesco Nastasi[1] · Giacomo Bergamini[2] · Vincenzo Balzani[2]

[1]Dipartimento di Chimica Inorganica, Chimica Analitica e Chimica Fisica, Università di Messina, Via Sperone 31, 98166 Messina, Italy
campagna@unime.it

[2]Dipartimento di Chimica "G. Ciamician", Università di Bologna, Via Selmi 2, 40126 Bologna, Italy

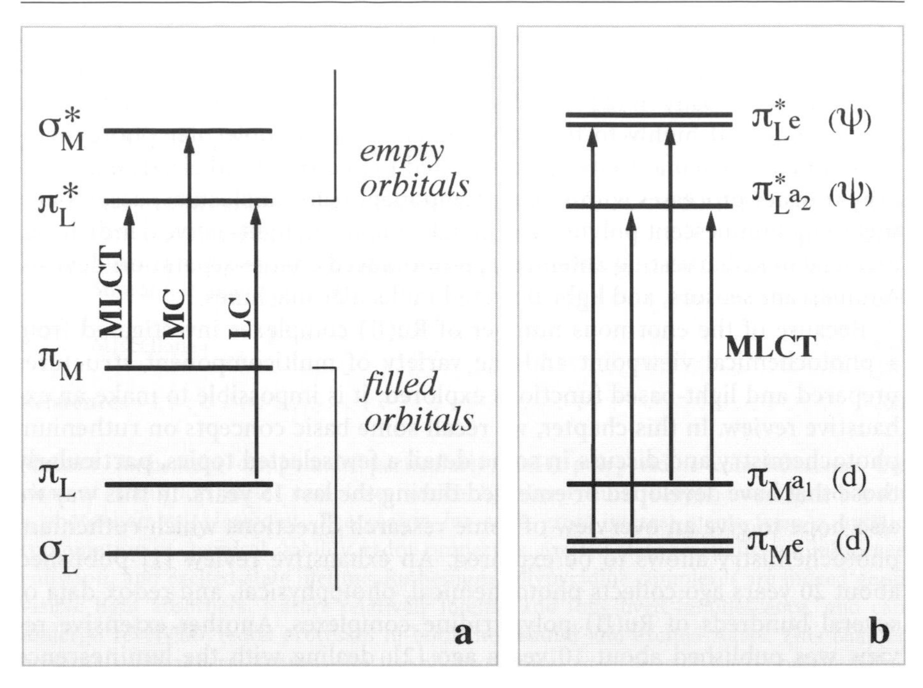

Fig. 1 **a** Simplified molecular orbital diagram for Ru(II) polypyridine complexes in octahedral symmetry showing the three types of electronic transitions occurring at low energies. **b** Detailed representation of the MLCT transition in D_3 symmetry

bitals gives rise to metal-centered (MC) excited states. Ligand-centered (LC) excited states can be obtained by promoting an electron from π_L to π_L^*. All these excited states may have singlet or triplet multiplicity, although spin-orbit coupling causes large singlet-triplet mixing, particularly in MC and MLCT excited states [6, 9–11].

The prototype $[Ru(bpy)_3]^{2+}$ (Fig. 2), as well as most of the $Ru(LL)_3{}^{2+}$ complexes (LL = bidentate polypyridine ligand), exhibits a D_3 symmetry [12]. Following Orgel's notation [13], the π^* orbitals may be symmetrical (χ) or

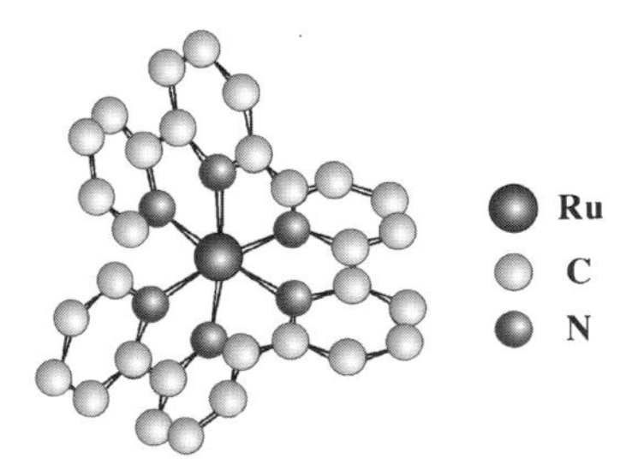

Fig. 2 Molecular structural formula of $[Ru(bpy)_3]^{2+}$

antisymmetrical (Ψ) with respect to rotation around the C_2 axis retained by each Ru(bpy) unit. A more detailed picture of the highest occupied molecular orbitals (HOMOs) and lowest unoccupied molecular orbitals (LUMOs) is shown in Fig. 1b [14–16]. The HOMOs are $\pi_M a_1(d)$ and $\pi_M e(d)$, which are mainly localized on the metal; the LUMOs are $\pi_L^* a_2(\Psi)$ and $\pi_L^* e(\Psi)$, which are mainly localized on the ligands. The ground state of the complex is a singlet, derived from the $\pi_M e(d)^4\ \pi_M a_1(d)^2$ electronic configuration.

According to Kasha's rule, only the lowest excited state and the upper states that can be populated on the basis of the Boltzmann equilibrium distribution may play a role in determining the photochemical and photophysical properties. The MC excited states of d^6 octahedral complexes are strongly displaced with respect to the ground-state geometry along metal–ligand vibration coordinates [17, 18].

When the lowest excited state is MC, it undergoes fast radiationless deactivation to the ground state and/or ligand dissociation reactions (Fig. 3). As a consequence, at room temperature the excited-state lifetime is very short, no luminescence emission can be observed [19], and very rarely bimolecular (or supramolecular) reactions can take place. LC and MLCT excited states are usually not strongly displaced compared to the ground-state geometry. Thus, when the lowest excited state is LC or MLCT (Fig. 3) it does not undergo fast radiationless decay to the ground state and luminescence can usually be observed. The radiative deactivation rate constant is somewhat higher for ^{3}MLCT than for ^{3}LC because of the larger spin–orbit coupling effect. For this reason, the ^{3}LC excited states are longer lived at low temperature in a rigid matrix and the ^{3}MLCT excited states are more likely to exhibit luminescence at room temperature in fluid solution.

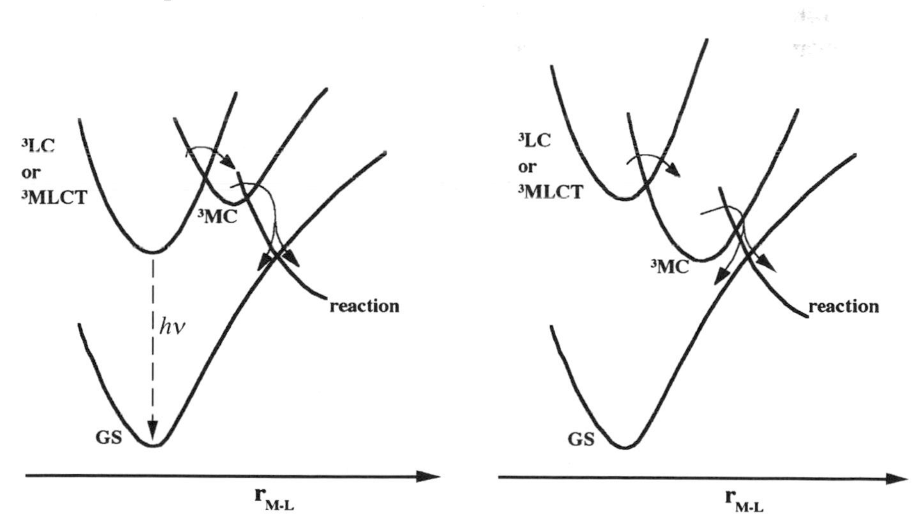

Fig. 3 Schematic representation of two limiting cases for the relative positions of ^{3}MC and ^{3}LC (or ^{3}MLCT) excited states

From the above discussion, it is clear that the excited-state properties of a complex are related to the energy ordering of its low-energy excited states and, particularly, to the orbital nature of its lowest excited state. The energy positions of the MC, MLCT, and LC excited states depend on the ligand field strength, the redox properties of metal and ligands, and intrinsic properties of the ligands, respectively [1, 2, 6]. Thus, in a series of complexes of the same metal ion, the energy ordering of the various excited states, and particularly the orbital nature of the lowest excited state, can be controlled by the choice of suitable ligands [1, 2, 5, 6]. It is therefore possible to design complexes having, at least to a certain degree, desired properties.

For most Ru(II) polypyridine complexes, the lowest excited state is a ^{3}MLCT level (or, better, a cluster [6] of closely spaced ^{3}MLCT levels, see later) which undergoes relatively slow radiationless transitions and thus exhibits relatively long lifetime and intense luminescence emission. Such a state is obtained by promoting an electron from a metal π_M orbital to a ligand π_L^* orbital (Fig. 1). The same π_L^* orbital is usually involved in the one-electron reduction process. For a long time it has been discussed whether in homoleptic complexes the emitting ^{3}MLCT state is best described with a multichelate ring-delocalized orbital (Fig. 4a) or a single chelate ring-localized orbital with a small amount of interligand interaction (Fig. 4b) [20]. This problem has been tackled with a variety of techniques on both reduced and excited complexes. Compelling evidence for "spatially isolated" [21] redox orbitals has been obtained from low-temperature cyclic voltammetry [22, 23], electron spin resonance [24], electronic absorption spectra of reduced species [25, 26], nuclear magnetic resonance [27], resonance Raman spectra [28, 29], and time-resolved infrared spectroscopy [30]. In the last 10 years, with the coming into play of ultrafast spectroscopic techniques, it has also been possible to investigate the nature of the Franck–Condon state and the rate constants of the localization/delocalization processes, as well as the interligand hopping (sometimes called "randomization of the excitation") in the MLCT excited state. These issues will be discussed in more detail later.

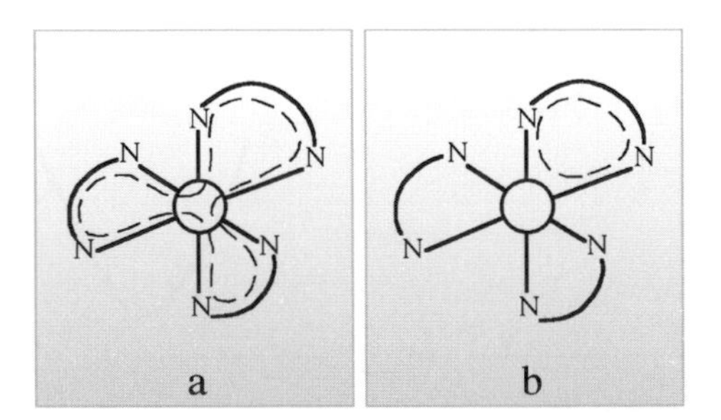

Fig. 4 Pictorial description of the electron promoted to the π_L^* orbital: **a** multichelate ring-delocalized orbital; **b** single chelate ring-localized orbital

3 $[Ru(bpy)_3]^{2+}$: The Prototype

To discuss the general properties of Ru(II) polypyridine complexes, it is convenient to refer to the properties of the prototype of this class of compounds, that is, $[Ru(bpy)_3]^{2+}$.

3.1 Absorption Spectrum

The absorption spectrum of $[Ru(bpy)_3]^{2+}$ is shown in Fig. 5 along with the proposed assignments [1, 4, 6, 14–16, 31]. The bands at 185 nm (not shown in the figure) and 285 nm have been assigned to spin-allowed LC $\pi \rightarrow \pi^*$ transitions by comparison with the spectrum of protonated bipyridine [32]. The two remaining intense bands at 240 and 450 nm have been assigned to spin-allowed MLCT $d \rightarrow \pi^*$ transitions. The shoulders at 322 and 344 nm might be MC transitions. In the long-wavelength tail of the absorption spectrum a shoulder is present at about 550 nm ($\varepsilon \sim 600\ M^{-1}\ cm^{-1}$) in an ethanol–methanol glass at 77 K [33]. This absorption feature is thought to be due to spin-forbidden MLCT transition(s).

In spite of the presence of the heavy Ru atom, it has been established that it is reasonable to assign the electronic transitions of $[Ru(bpy)_3]^{2+}$ as being due to "singlet" or "triplet" states. In particular, a singlet character $\leq 10\%$ has been estimated [10, 34] for the lowest-lying excited states of $[Ru(bpy)_3]^{2+}$. The maximum of the ^{1}MLCT band at ~ 450 nm is slightly sensitive to solvent, suggesting an instantaneous sensing of the formation of the dipolar excited-state $[Ru^{3+}(bpy)_2(bpy)^-]^{2+}$ [35].

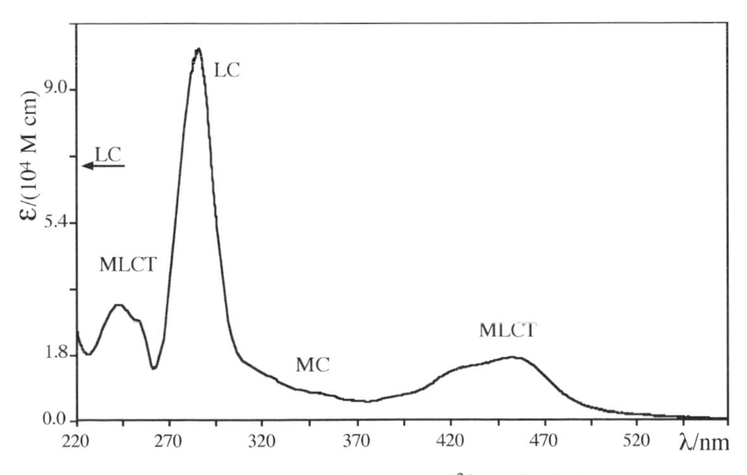

Fig. 5 Electronic absorption spectrum of $[Ru(bpy)_3]^{2+}$ in alcoholic solution

With increasing temperature, the emission lifetime (Fig. 8) and quantum yield decrease [1, 4, 6, 8, 32, 61–79]. This behavior may be accounted for by a stepwise term and two Arrhenius terms [1–3]:

$$1/\tau = k_0 + \frac{B}{1 + \exp[C(1/T - 1/T_B)]} + A_1 \exp(-\Delta E_1/RT) + A_2 \exp(-\Delta E_2/RT) \,. \quad (1)$$

The value of the various parameters is somewhat dependent on the nature of the solvent. In propionitrile–butyronitrile (4 : 5 v/v) the values are as follows [70, 71]: $k_0 = 2 \times 10^5\ s^{-1}$; $B = 2.1 \times 10^5\ s^{-1}$; $A_1 = 5.6 \times 10^5\ s^{-1}$; $\Delta E_1 = 90\ cm^{-1}$; $A_2 = 1.3 \times 10^{14}\ s^{-1}$; $\Delta E_2 = 3960\ cm^{-1}$. Included in k_0 are the radiative k_0(r) and nonradiative k_0(nr) rate constants at 84 K. The stepwise term B is due to the melting of the matrix (100–150 K) and corresponds to the coming into play of vibrations capable of facilitating radiationless deactivation [8, 71]. In the same temperature range a red shift of $\sim 1000\ cm^{-1}$ is observed in the maximum of the emission band, and it is mainly attributed to reorganization of solvent molecules around the excited state in fluid solution before emission takes place [8, 71]. The Arrhenius term with $A_1 = 5.6 \times 10^5\ s^{-1}$ and $\Delta E_1 = 90\ cm^{-1}$ is thought to correspond to the thermal equilibration with a level lying at slightly higher energy and having the same electronic nature (so it would be a fourth MLCT state [6], considering the lowest-lying MLCT state is made of three sublevels as described before). The second Arrhenius term corresponds to a thermally activated surface crossing to an upper-lying ^{3}MC level which undergoes fast deactivation. Identification of this higher level as a ^{3}MC state is based upon the observed photosubstitution behavior at elevated temperatures [61], consistent with established photoreactivity patterns for d^6 metal complexes [17, 52].

Experiments carried out with $[Ru(bpy)_3]^{2+}$ and $[Ru(bpy\text{-}d_8)_3]^{2+}$ in H_2O and D_2O [61, 80, 81] indicate that k_0(nr) is sensitive to deuteration, as expected for a weak-coupled radiationless process [6, 82–84]. By contrast, A_2 is insensitive to deuteration, supporting a strong-coupled (surface crossing) deactivation pathway, which may be related to the observed photosensitivity. It should be noted that the decrease in lifetime on melting has also been explained on the basis of the energy gap law because of the corresponding red shift in the emission band [6].

Finally, it should be noted that at 77 K the emission spectrum of $[Ru(bpy)_3]^{2+}$, as well as that of most Ru(II) polypyridine complexes, exhibits a vibrational structure (see Fig. 7). This structure is assigned to the vibrational progression, and its energy spacing is about $1300\ cm^{-1}$, equivalent to the C – N and C – C stretching energy of the aromatic rings, thus indicating that such stretchings are the dominant accepting modes for deactivation of the ^{3}MLCT state.

3.4
Photosubstitution and Photoracemization Processes

Although $[Ru(bpy)_3]^{2+}$ is normally considered as photochemically inert toward ligand substitution, this is not strictly true [1, 6, 14, 15]. In aqueous solution the quantum yield of $[Ru(bpy)_3]^{2+}$ disappearance is in the range 10^{-5}–10^{-3}, depending on both the pH of the solution and temperature [1]. In chlorinated solvents such as CH_2Cl_2, the photochemistry of $[Ru(bpy)_3]X_2$ ($X = Cl^-$, Br^-, NCS^-) is well behaved [64, 85], giving rise to $[Ru(bpy)_2X_2]$ as the final product. The quantum yields are in the range 10^{-1}–10^{-2}. The PF_6^- salt is photoinert. A substantial difference between aqueous and CH_2Cl_2 solutions is that salts of $[Ru(bpy)_3]^{2+}$ are completely ion-paired in the latter medium.

A detailed mechanism for the ligand photosubstitution reaction of $[Ru(bpy)_2X_2]$ has been proposed [6, 64] (Fig. 9). According to this mechanism, thermally activated formation of a ^{3}MC excited state (vide supra) leads to the cleavage of a Ru – N bond, with formation of a five-coordinate square pyramidal species. In the absence of coordinating ions, as with the PF_6^- salt, this square pyramidal species returns to $[Ru(bpy)_3]^{2+}$. When coordinating anions are present, as in the Cl^- salt, a hexacoordinated monodentate bpy intermediate is formed. Once formed, this monodentate bpy species can undergo loss of bpy and formation of $[Ru(bpy)_2X_2]$, or a "self-annealing" process (chelate ring closure), with re-formation of $[Ru(bpy)_3]^{2+}$. The "self-annealing" protective step is favored in aqueous solution, presumably because

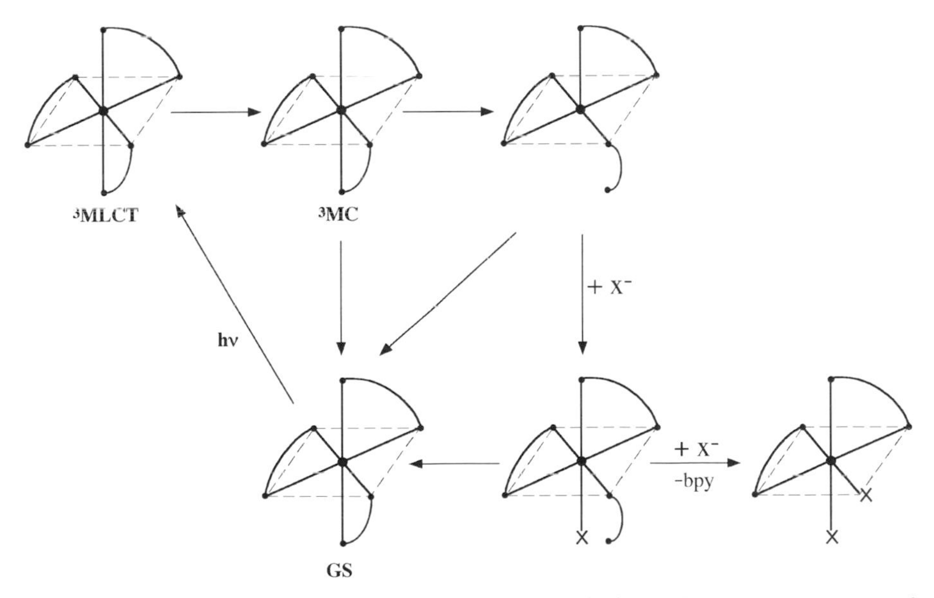

Fig. 9 Scheme of the proposed mechanism for ligand photosubstitution reactions of $[Ru(bpy)_3]X_2$

trum of Q^- to monitor product formation. By contrast, $[Ru(bpy)_3]^+$ exhibits a strong absorption band at 510 nm, which is particularly useful to investigate reductive quenching reactions.

Following some pioneering works [96–100], literally hundreds of bimolecular excited-state reactions of $[Ru(bpy)_3]^{2+}$ and of its derivatives have been studied [101]. Here we only illustrate a few examples to show that these excited-state reactions can be used for mechanistic studies as well as for potential applications of the greatest interest.

The early interest in $[Ru(bpy)_3]^{2+}$ photochemistry arose from the possibility of using its long-lived excited state as energy donor in energy transfer processes. Although several sensitized reactions attributed to energy transfer processes (see, e.g., [102, 103]) were later shown to proceed via electron transfer [100], there are some very interesting cases in which energy transfer has been firmly demonstrated. A clear example is the quenching of $^*[Ru(bpy)_3]^{2+}$ by $[Cr(CN)_6]^{3-}$, where sensitized phosphorescence of the chromium complex has been observed both in fluid solution [104–108] and in the solid state [109–111]:

$$^*[Ru(bpy)_3]^{2+} + [Cr(CN)_6]^{3-} \rightarrow [Ru(bpy)_3]^{2+} + (^2E_g)[Cr(CN)_6]^{3-} \tag{5}$$

$$(^2E_g)[Cr(CN)_6]^{3-} \rightarrow [Cr(CN)_6]^{3-} + h\nu \,. \tag{6}$$

Energy transfer from $^*[Ru(bpy)_3]^{2+}$ to $[Cr(CN)_6]^{3-}$ was also used to demonstrate that the photosolvation reaction observed upon direct excitation of $[Cr(CN)_6]^{3-}$ does not originate from the luminescent ^2E_g state of the chromium complex [104, 112].

It should be pointed out that both reductive and oxidative $^*[Ru(bpy)_3]^{2+}$ electron transfer quenchings by $[Cr(CN)_6]^{3-}$ are thermodynamically forbidden because it is very difficult to reduce or oxidize $[Cr(CN)_6]^{3-}$ [108]. $[Cr(bpy)_3]^{3+}$, by contrast, can be very easily reduced and with this quencher oxidative electron transfer prevails over energy transfer

$$^*[Ru(bpy)_3]^{2+} + [Cr(bpy)_3]^{3+} \xrightarrow{k=3.3\times10^9\ M^{-1}\ s^{-1}} [Ru(bpy)_3]^{3+} + [Cr(bpy)_3]^{2+} \,, \tag{7}$$

as is shown by the appearance of the $[Cr(bpy)_3]^{2+}$ absorption spectrum in flash photolysis experiments [113]. Equation 7 converts 71% of the spectroscopic energy (2.12 eV) of the excited-state reactant into chemical energy of the products. As usually happens in these simple homogeneous systems, the converted energy cannot be stored but is immediately dissipated into heat by the back electron transfer reaction:

$$[Ru(bpy)_3]^{3+} + [Cr(bpy)_3]^{2+} \xrightarrow{k=2\times10^9\ M^{-1}\ s^{-1}} [Ru(bpy)_3]^{2+} + [Cr(bpy)_3]^{3+} \,. \tag{8}$$

A carefully studied example of reductive electron transfer quenching (Eq. 9) is that involving Eu^{2+}_{aq} as a quencher [114, 115]:

$$^{*}[Ru(bpy)_3]^{2+} + Eu_{aq}{}^{2+} \xrightarrow{k=2.8\times10^7\ M^{-1}\ s^{-1}} [Ru(bpy)_3]^{+} + Eu_{aq}{}^{3+} . \qquad (9)$$

The difference spectrum obtained by flash photolysis after a 30-ns light pulse shows a bleaching in the region around 430 nm due to depletion of $[Ru(bpy)_3]^{2+}$ and an increased absorption around 500 nm due to the formation of $[Ru(bpy)_3]^{+}$ (note that both Eu^{2+}_{aq} and Eu^{3+}_{aq} are transparent in this spectral region). Clear kinetic evidence for reductive quenching comes from the observation that the growth of the absorption at 500 nm occurs at a rate equal to the rate of decay of the luminescence emission of $^{*}[Ru(bpy)_3]^{2+}$. As it may happen in excited-state reactions, the products of Eq. 9 have a high energy content and thus they give rise to a back electron transfer reaction

$$[Ru(bpy)_3]^{+} + Eu_{aq}{}^{3+} \xrightarrow{k=2.7\times10^7\ M^{-1}\ s^{-1}} [Ru(bpy)_3]^{2+} + Eu_{aq}{}^{2+} , \qquad (10)$$

which can be monitored (on a longer timescale) through the recovery of the 430-nm absorption or the disappearance of the 500-nm absorption.

In several cases direct evidence for energy transfer quenching (i.e., sensitized luminescence or absorption spectrum of the excited acceptor) or electron transfer quenching (i.e., absorption spectrum of redox products) is difficult or even impossible to obtain for bimolecular processes. In such cases, free energy correlations of rate constants are quite useful to elucidate the reaction mechanism [108, 116–118]. As we will see later, photoinduced energy and electron transfer processes can take place very easily in suitably organized supramolecular systems.

3.6 Chemiluminescence and Electrochemiluminescence Processes

As mentioned in the introductory chapter (Balzani et al. 2007, in this volume) [119], excited states can be generated in very exergonic electron transfer reactions. Formation of excited states can be easily demonstrated when the excited states are luminescent species. Because of its stability in the reduced and oxidized forms and the strong luminescence of its excited state, $[Ru(bpy)_3]^{2+}$ is an extremely versatile reactant for a variety of chemiluminescent processes [32, 120–124].

In principle, there are two ways to generate the luminescent $^{*}[Ru(bpy)_3]^{2+}$ excited state in chemical reactions. One way (Eq. 11) is to oxidize $[Ru(bpy)_3]^{+}$ with a species X having reduction potential $E^0(X/X^-)$ more positive than 0.84 V, and another way (Eq. 12) is to reduce $[Ru(bpy)_3]^{3+}$ with a species Y^-

whose potential $E^0(Y/Y^-)$ is more negative than – 0.86 V (see also Fig. 10).

$$[Ru(bpy)_3]^+ + X \rightarrow {}^*[Ru(bpy)_3]^{2+} + X^- \quad (11)$$

$$[Ru(bpy)_3]^{3+} + Y^- \rightarrow {}^*[Ru(bpy)_3]^{2+} + Y \quad (12)$$

$${}^*[Ru(bpy)_3]^{2+} \rightarrow [Ru(bpy)_3]^{2+} + h\nu\,. \quad (13)$$

A variety of oxidants (e.g., $S_2O_8{}^{2-}$ [125, 126]) and reductants (e.g., e^-_{aq} [127], hydrazine and hydroxyl anion [128], oxalate ion [129, 130]) have been used in these chemiluminescent processes. In some cases (e.g., with OH^-), the reaction mechanism cannot be a simple outer sphere electron transfer reaction and the emitting species could be a slightly modified (on the ligands) complex. It should also be pointed out that minor amounts of oxidizing and reducing impurities are sufficient to produce luminescence in chemiluminescence and electrochemiluminescence experiments [131].

The most interesting way [132] to obtain chemiluminescence from $[Ru(bpy)_3]^{2+}$ solutions is probably to produce the oxidized and/or reduced form of the complex "in situ" by electrochemical methods. Three classical experiments of this type can be performed:

(a) To pulse the potential applied to a working electrode between the oxidation and reduction potentials of $[Ru(bpy)_3]^{2+}$ in a suitable solvent [132, 133]. In such a way the reduced and oxidized forms produced in the same region of space can undergo a comproportionation reaction where enough energy is available to produce an excited state and a ground state (see also Fig. 10):

$$[Ru(bpy)_3]^{2+} + e^- \rightarrow [Ru(bpy)_3]^+ \quad (14)$$

$$[Ru(bpy)_3]^{2+} - e^- \rightarrow [Ru(bpy)_3]^{3+} \quad (15)$$

$$[Ru(bpy)_3]^{3+} + [Ru(bpy)_3]^+ \rightarrow {}^*[Ru(bpy)_3]^{2+} + [Ru(bpy)_3]^{2+}\,. \quad (16)$$

(b) To reduce $[Ru(bpy)_3]^{2+}$ in the presence of a strong oxidant (reductive oxidation). For example, luminescence is obtained upon continuous reduction of $[Ru(bpy)_3]^{2+}$ at a working electrode in the presence of $S_2O_8{}^{2-}$ [125, 126]. This oxidant in a first one-electron oxidation reaction generates the very powerful oxidant $SO_4{}^-$ that can either oxidize $[Ru(bpy)_3]^+$ to ${}^*[Ru(bpy)_3]^{2+}$ (Eq. 18) or $[Ru(bpy)_3]^{2+}$ to $[Ru(bpy)_3]^{3+}$ (Eq. 19), which then reacts with $[Ru(bpy)_3]^+$ (Eq. 16) to yield the luminescent excited state:

$$[Ru(bpy)_3]^{2+} + e^- \rightarrow [Ru(bpy)_3]^+ \quad (14)$$

$$[Ru(bpy)_3]^+ + S_2O_8{}^{2-} \rightarrow [Ru(bpy)_3]^{2+} + SO_4{}^- + SO_4{}^{2-} \quad (17)$$

$$[Ru(bpy)_3]^+ + SO_4{}^- \rightarrow {}^*[Ru(bpy)_3]^{2+} + SO_4{}^{2-} \quad (18)$$

$$[Ru(bpy)_3]^{2+} + SO_4{}^- \rightarrow [Ru(bpy)_3]^{3+} + SO_4{}^{2-} \quad (19)$$

$$[Ru(bpy)_3]^+ + [Ru(bpy)_3]^{3+} \rightarrow {}^*[Ru(bpy)_3]^{2+} + [Ru(bpy)_3]^{2+}\,. \quad (16)$$

(c) To oxidize $[Ru(bpy)_3]^{2+}$ in the presence of a strong reductant (oxidative reduction). For example, light is generated upon continuous oxidation of $[Ru(bpy)_3]^{2+}$ at a working electrode in the presence of $C_2O_4{}^{2-}$ [129, 130]. This reductant in a first one-electron reaction generates the strongly reducing $CO_2{}^-$ radical that can reduce $[Ru(bpy)_3]^{3+}$ to the excited $*[Ru(bpy)_3]^{2+}$

$$[Ru(bpy)_3]^{2+} - e^- \rightarrow [Ru(bpy)_3]^{3+} \tag{15}$$

$$[Ru(bpy)_3]^{3+} + C_2O_4{}^{2-} \rightarrow [Ru(bpy)_3]^{2+} + CO_2 + CO_2{}^- \tag{20}$$

$$[Ru(bpy)_3]^{3+} + CO_2{}^- \rightarrow *[Ru(bpy)_3]^{2+} + CO_2 \,. \tag{21}$$

These chemiluminescent electron transfer reactions are quite interesting from an applicative [134–136] as well as from a theoretical viewpoint. Actually, method a is at the basis of electroluminescent materials, such as organic light-emitting diodes (OLEDs) and similar devices, which are receiving increasing interest for practical applications [137–141].

4 Some Important Features of Ru(II) Polypyridine Complexes

4.1 Nonradiative Decay Rate Constants and Emission Spectral Profiles of Ru(II) Polypyridine Complexes

Radiationless decay from MLCT states of metal polypyridine complexes occurs with energy release into medium-frequency (polypyridyl-based) modes and, to a lower degree, low-frequency modes and solvent [4, 142–149]. Averaging the medium-frequency modes which mainly promote the transition and combining low-frequency modes, including solvent, into a single mode, treated classically, the rate constant for radiationless decay k_{nr} is predicted to follow the so-called energy gap law [150–154]. Most of the work to define this topic has been made by using Ru(II) polypyridine complexes as models; however, the approach also applies to any MLCT emitter, as largely demonstrated for Os(II) [146, 147, 155] and Re(I) polypyridine [147, 149, 156] complexes. Actually, the energy gap law can be expressed by Eq. 22, where β_0 includes the vibrationally induced electronic matrix element and F(calc) is the vibrational overlap factor (the quantity 1s in Eq. 22 is used to give unitless expression):

$$\ln(k_{nr} \cdot 1s) = \ln \beta_0 + \ln[F(\text{calc})] \,. \tag{22}$$

In a simplified version, F(calc) can be expressed as in Eq. 23 [157]:

$$F(\text{calc}) \propto \left(\frac{-\gamma E_0}{\hbar\omega}\right) \quad \gamma = \ln\left(\frac{E_0}{S_M \hbar\omega}\right) - 1 \,. \tag{23}$$

However, some experimental data contrast with the scheme shown in Fig. 11. For example, a time-resolved resonance Raman study indicates that in $[Ru(bpy)_3]^{2+}$ even the initial excitation is localized [172]. Moreover, the recently reported femtosecond transient absorption spectrum of $[Ru(bpy)_3]^{2+}$ in the UV region suggests that complete relaxation within the emitting triplet MLCT state takes several tens of picoseconds, and that randomization of triplet MLCT is complete in less than 500 fs [173]. If this latter point is correct, interligand hopping should largely occur from nonrelaxed states, probably even partly in the singlet state. In the relaxed triplet MLCT state, interligand hopping could be slower, but it would be difficult to measure since randomization would already have happened.

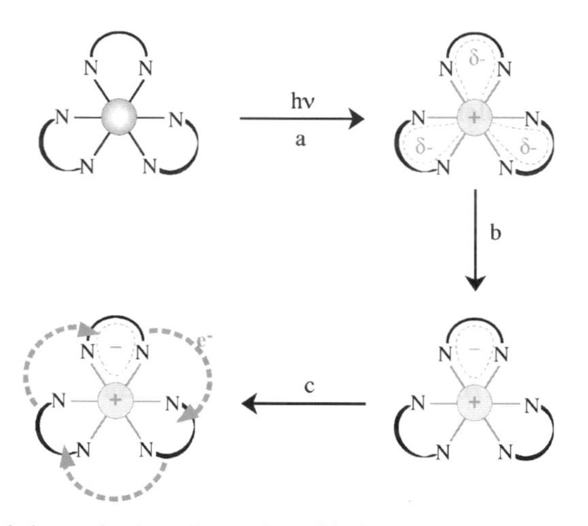

Fig. 11 Picture of the early-time dynamics of light excitation in the MLCT singlet of $[Ru(bpy)_3]^{2+}$. A delocalizated Franck–Condon state is formed (**a**), which becomes localized on a single ligand (**b**) and then becomes "randomized" by interligand hopping (**c**)

Related to the excited-state dynamics at short times after excitation, broadband femtosecond fluorescence spectroscopy of $[Ru(bpy)_3]^{2+}$ has been recently reported, as already mentioned [171]. The authors get 15 ± 10 fs as the lifetime for the singlet emission, which is centered at about 520 nm.

4.3 Ru(II) Complexes Based on Tridentate Polypyridine Ligands

An important family of Ru(II) polypyridine complexes is that based on tridentate ligands, with $[Ru(terpy)_2]^{2+}$ as a prototype (terpy = 2,2′ : 6′,2″-terpyridine). The absorption, emission, and redox properties of $[Ru(terpy)_2]^{2+}$ are similar to those of $[Ru(bpy)_3]^{2+}$, except that $[Ru(terpy)_2]^{2+}$ is essentially nonluminescent at room temperature, with a lifetime of the ^{3}MLCT state in degassed acetonitrile at room temperature of about 250 ps (measured by transient absorption spectroscopy [3]), compared with a value of about 1 μs exhibited by $[Ru(bpy)_3]^{2+}$ under the same conditions [1]. Such a short excited-state lifetime is very disappointing, as $[Ru(terpy)_2]^{2+}$ has some advantage over $[Ru(bpy)_3]^{2+}$ from a structural point of view. Whereas $[Ru(bpy)_3]^{2+}$ can exist as a mixture of Λ and Δ isomers, and the isomer problem can become even more complicated for polynuclear species based on "asymmetric" bidentate ligands such as 2,3-bis(2′-pyridyl)pyrazine (2,3-dpp), $[Ru(bpy)_3]^{2+}$ is achiral. Moreover, by taking advantage of *para* substituents on the central pyridine of the terpy ligand, $[Ru(terpy)_2]^{2+}$ can give rise to supramolecular architectures perfectly characterized from a structural viewpoint, in particular to multinuclear one-dimensional ("wire"-like) species. The reason for the poor photophysical properties of Ru(II) complexes with tridentate polypyridine ligands at room temperature, compared to Ru(II) species with bidentate chelating polypyridine, stems from the bite angle of the tridentate ligand that leads to a weaker ligand field strength and thus to lower-energy MC states as compared to Ru(II) complexes of bpy. The thermally activated process from the potentially emitting ^{3}MLCT state to the higher-lying ^{3}MC state is therefore more efficient in $[Ru(terpy)_2]^{2+}$ and its derivatives and leads to fast deactivation of the excited state by nonradiative processes [1, 3, 4], although terpy-type Ru complexes are inherently more photostable than bpy-type ones because of a stronger chelating effect.

Much effort has been devoted to the design and synthesis of tridentate polypyridine ligands, leading to Ru(II) complexes with improved photophysical properties [3, 78, 92, 174–181]. For example, the use of ligands containing electron-withdrawing and -donor substituents on tpy increases the gap between the ^{3}MLCT and the ^{3}MC states [174]. An increase in such an energy gap has also been obtained by the use of cyclometallating ligands [177]. Unavoidably, the stabilization of ^{3}MLCT states causes an increase of the rate constant for radiationless decay to the ground state. This latter effect can be balanced by extension of the π^* orbital by appropriate substituents, which increases the delocalization of the acceptor ligand of the MLCT excited state leading to a smaller Franck–Condon factor for nonradiative decay [78, 175, 176, 178, 179, 182–188]. In this regard, species based on ethynyl-substituted terpy ligands feature particularly interesting photophysical properties [175, 176, 182]. Various approaches to improve the photophysical properties of Ru(II) complexes with tridentate polypyridine ligands have been reviewed [175, 182].

The bis-tridentate Ru(II) polypyridine complex with the best photophysical properties reported up to now is probably the species **1**, based on the 2,6-bis(8′-quinolinyl)pyridine ligand [189]. This species exhibits ^{3}MLCT emission with a maximum at 700 nm, with a lifetime of 3.0 μs and a quantum yield of 0.02 in deoxygenated methanol–ethanol solution at room temperature. The emission maximum blue-shifts to 673 nm at 77 K in the same solvent mixture, exhibiting a luminescence lifetime of 8.5 μs and a quantum

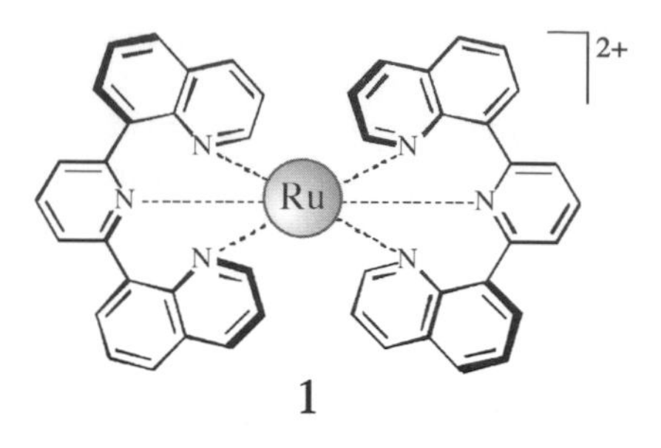

yield of 0.06. The authors attribute these excellent (particularly at room temperature) photophysical properties to the relief of structural distortion from the ideal octahedral geometry, due to ligand design. Actually, X-ray characterization of the compound reveals a quasi-ideal octahedral geometry around the metal center.

4.4 Interplay Between Multiple Low-Lying MLCT States Involving a Single Polypyridine Ligand

Usually, there is a linear relationship between redox data, namely first oxidation and reduction potentials of Ru complexes, and spectroscopic parameters such as MLCT absorption and emission bands, provided that the considered compounds constitute a homogeneous series [1, 4, 190–192]. This relationship is based on the fact that the orbitals involved in metal-based oxidation and ligand-based reduction processes are the same (to a first approximation) as those involved in the MLCT absorption and emission transitions. Differences in solvent effects for redox and spectroscopic processes should be constant, so that the relationship is still linear, although the slops are not unitary [1, 191]. However, whereas until 20 years ago this relationship was followed by almost all the Ru complexes reported at that time, and exceptions were rare [193] and partly unexplained, Ru complexes which do not follow the rule, in particular as far as the absorption spectra are concerned, have become quite common in recent years. The availability of several examples allowed the development of a general interpretation of this behavior. In all the cases that do not obey the linear relationship, ligands characterized by a large aromatic framework are present.

It is now clear that the apparent mismatched relationship is linked to the presence of multiple low-energy MLCT transitions to a single polypyridine ligand, with one such transition being essentially almost invisible spectroscopically. The key feature here is that the "single" polypyridine ligand can actually be viewed as being made of two "separated" subunits (Fig. 12) with the LUMO centered on a part of the ligand framework which is not significantly coupled with the metal-based HOMOs (in other words, the LUMO does

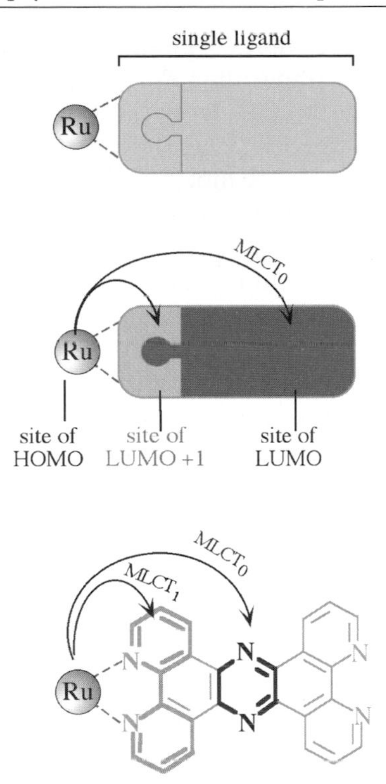

Fig. 12 Structural formula of two possible MLCT transitions in ruthenium complexes with large polypyridine ligands

not receive a significant contribution from the chelating nitrogens). In these systems, the lowest-energy MLCT transition ($MLCT_0$) can have vanishing oscillator strength and thus it does not significantly contribute to the absorption spectrum. On the contrary, a closely lying LUMO+1 (centered on a different moiety of the large ligand) receives a significant contribution from the chelating nitrogens, so it is largely coupled with the metal-based HOMO(s); as a consequence, its corresponding MLCT transition (the $MLCT_1$ transition) dominates the absorption spectrum. Since reduction takes place in the LUMO, the linear relationship between absorption spectra and redox potential cannot be followed. This case will also be discussed in Sects. 5 and 6, for specific systems.

As far as the relationship between emission spectra and redox potentials is concerned, whether it is followed or not depends on how fast the interconversion between the $MLCT_1$ and $MLCT_0$ states is, compared to the intrinsic decay of the $MLCT_1$ excited state (here it is assumed that $MLCT_0$ is lower in energy than $MLCT_1$; otherwise, the relationship is always followed, except for very particular cases). Solvent, temperature, driving force, and medium effects are very important in this regard. For example, at room temperature in fluid solution the mononuclear complex $[(phen)_2Ru(tpphz)]^{2+}$ (**2**, phen = 1,10-phenanthroline; tpphz = tetrapyrido[3,2-*a*:2′,3′-*c*:3″,2″-*h*:2″,3″-*j*]phenazine) exhibits emission from its $MLCT_1$ level (λ_{max} = 625 nm, τ = 1.25 ms, Φ = 0.07), while the dinuclear species $[(phen)_2Ru(tpphz)Ru(phen)_2]^{4+}$ (**3**) emits from its $MLCT_0$ level (λ_{max} = 710 nm, τ = 0.100 ms, Φ = 0.005) [194]. The absorption spectra of both compounds in the visible region are very similar to one another (apart from the intensity), with the lowest-energy MLCT band maximizing at about 440 nm in both cases. In a rigid matrix at 77 K, both the mononuclear and dinuclear metal complexes exhibit emission at about 585 nm (lifetime in the microsecond timescale), typical of the $MLCT_1$ level. Such results are interpreted on considering that the ligand tpphz has two empty orbitals close in energy: the LUMO is centered on the central pyrazine, with negligible contribution from the chelating nitrogen atoms, and the LUMO+1 is essentially a bpy-type orbital. Reduction potential data of the complexes indicate that LUMO+1 of tpphz is hardly affected on passing from mononuclear to dinuclear species, whereas the LUMO of tpphz is stabilized. For both $[(phen)_2Ru(tpphz)]^{2+}$ and $[(phen)_2Ru(tpphz)Ru(phen)_2]^{4+}$, the absorption spectrum is dominated by Ru-to-$tpphz_{LUMO+1}$ charge transfer (i.e., $MLCT_1$) transition—almost coincident to the Ru-to-phen charge transfer transition—which occurs at roughly the same energy in mononuclear and dinuclear species, with the Ru-to-$tpphz_{LUMO}$ charge transfer (i.e., $MLCT_0$) transition not contributing to the absorption feature. Because of the different stabilization of $MLCT_1$ and $MLCT_0$ on passing from mononuclear to dinuclear species (see above), the driving force for the $MLCT_1$-to-$MLCT_0$ interconversion is more favorable, and therefore faster, in the dinuclear species. As a consequence, $MLCT_1$-to-$MLCT_0$ decay does not compete with the direct decay of $MLCT_1$ to the ground state in the mononuclear species, whereas it is

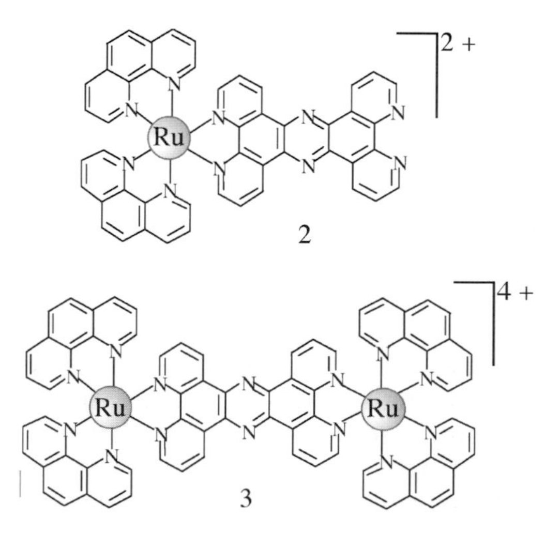

faster and efficient in the dinuclear species. However, at 77 K the $MLCT_1$-to-$MLCT_0$ interconversion, which cannot occur without solvent reorganization, becomes inefficient in both systems. Essentially the same experimental behavior is featured by many other Ru(II) polypyridine complexes, and can be interpreted in a similar way [195–205].

Time-resolved transient absorption spectroscopy [206, 207] confirmed that the $MLCT_1$-to-$MLCT_0$ excited-state conversion in $[(phen)_2Ru(tpphz)Ru(phen)_2]^{4+}$ at room temperature is solvent dependent. Indeed, it was 120 ps in dichloromethane and faster than 40 ps in acetonitrile [206, 207]. The solvent dependence can be attributed to the difference in the $MLCT_1/MLCT_0$ energy gap in the two solvents. It has to be considered, in fact, that the "charge separation" between donor and acceptor orbitals, and, as a consequence, the coulombic stabilization, is quite different in the two types of MLCT states. A nonnegligible reorganization energy is therefore expected for the $MLCT_1$-to-$MLCT_0$ transition. A detailed study of the temperature dependence of the luminescence properties of species exhibiting this interesting behavior would be quite useful, but to our knowledge it has not yet been reported.

5 Ruthenium and Supramolecular Photochemistry

Supramolecular photochemistry has played a prominent role in chemical research since its definition in the late 1980s [208, 209]. The operational definition of supramolecular species is discussed (Balzani et al. 2007, in this volume) [119], so it will not be further commented on here. Since Ru(II) polypyridine complexes exhibit very interesting photochemical properties and can be prepared by relatively easy synthetic methods, even with made-to-order properties, the number of photoactive supramolecular species based on Ru(II) complexes has rapidly become extraordinarily large. Supramolecular systems in which donor and acceptor units are placed at designed distances can undergo photoinduced energy and electron transfer process (first-order kinetics) even in the case of short-lived excited states [208, 209].

Indeed, Ru(II) polypyridine compounds have been extensively used as photoactive units in supramolecular systems either exclusively made of metal-based components, such as molecular racks, grids, and dendrimers, or in systems whose other active components of the assemblies are of an organic nature. In both cases, the final goals of the supramolecular systems are essentially two, reflecting the nature of the whole of photochemical science: (1) systems designed for the conversion of light energy into other forms of energy, essentially chemical energy or electricity; and (2) systems focused on the elaboration of the information, including sensors. Quite often these two aspects are intertwined; for example, long-range photoinduced electron and energy transfer processes are important both for the elaboration of optical information signals and for light-harvesting systems.

5.1 Photoinduced Electron/Energy Transfer Across Molecular Bridges in Dinuclear Metal Complexes

Dinuclear metal complexes containing Ru(II) polypyridine subunits, where the metal centers are separated by molecular components (bridges), are particularly suited to investigating photoinduced electron and energy transfer processes, whose rate constants can give information on the electronic coupling mediated by the bridge. The latter topic has been recently reviewed and deeply discussed [210]. The number of photoactive (usually, luminescent) dinuclear metal complexes based on Ru(II) subunits is very large. The last exhaustive review dealing with such species was published about 10 years ago [2]. Today it is impossible to be exhaustive even in this relatively narrow field. Therefore, we will only present a few examples. In most cases, Ru(II) subunits, which play the role of donors, are coupled to Os(II) units, which play the role of acceptors in photoinduced energy transfer processes. In all cases, the dinuclear homometallic Ru(II) species have also been investigated for comparison purposes. Their photophysical properties can be found in the original references.

It is important to note that to have control of the distance between the metal centers, the bridges have to be rigid as much as possible. Therefore, it is not surprising that oligophenylenes have often been employed. In the series of dinuclear dyads **4** [211], having the general formula $[Ru(bpy)_3]^{2+}$-$(ph)_n$-(R_2ph)-$(ph)_n$-$[Os(bpy)_3]^{2+}$ (ph = 1,4-phenylene; $n = 1, 2, 3$), excitation of the $[Ru(bpy)_3]^{2+}$ unit is followed by energy transfer to the $[Os(bpy)_3]^{2+}$ unit, as shown by the sensitized emission of the latter. For the compound with $n = 3$, with a total of seven phenylene spacers, the rate constant k_{en} for energy transfer over the 4.2-nm metal-to-metal distance is $1.3 \times 10^6\ s^{-1}$ in acetonitrile solution at room temperature. This was probably the first example of

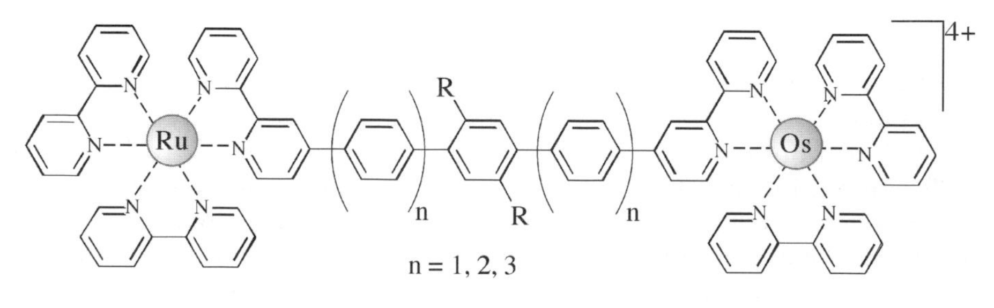

a systematic study on the distance dependence of energy transfer rates for Ru(II)–Os(II) dyads. A Dexter-type mechanism for the Ru(II)–Os(II) energy transfer was proposed, and an attenuation factor β of 0.32 Å^{-1} for photoinduced energy transfer was obtained from the ln k_{en} vs metal–metal distance plot.

In the $[Ru(bpy)_3]^{2+}$-$(ph)_n$-$[Os(bpy)_3]^{3+}$ compounds, obtained by chemical oxidation of the Os-based moiety, photoexcitation of the $[Ru(bpy)_3]^{2+}$ unit causes the transfer of an electron to the Os-based one with a rate constant (k_{el}) of 3.4×10^7 s^{-1} for $n = 3$. Unless the electron added to the $[Os(bpy)_3]^{3+}$ unit is rapidly removed, a back electron transfer reaction (rate constant 2.7×10^5 s^{-1} for $n = 3$) takes place from the $[Os(bpy)_3]^{2+}$ unit to the $[Ru(bpy)_3]^{3+}$ one [211]. The rate constants of all the transfer processes in the series of complexes decrease, as expected, with decreasing length of the oligophenylene spacer, whereas they were practically unaffected by temperature.

Interestingly, a series of analogous dyads missing the central substituted phenylene (**5**) was successively prepared [212, 213] and the results of the two series have been compared: for the dyads containing bridges made of a total of three and five phenylene spacers, the rate constants of photoinduced energy transfer are higher in the nonsubstituted phenyl series. This was attributed to effects of inter-phenylene twist angle on the electronic coupling between donor and acceptor subunits [213].

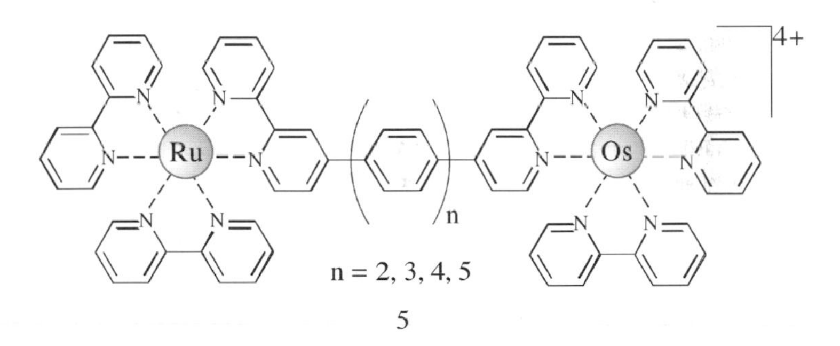

5

The presence of *meta* substitution in oligophenylene bridges versus the all-*para* systems have been evidenced by the dyads **6** [210, 213]. In these species, the photoinduced energy transfer rate constants are lower than the rates for the respective compounds containing all-*para* phenylene units: for example, for spacers made of three and five phenylenes, respectively, k_{en} is 1.32×10^9 and 6.67×10^7 s^{-1} for the *meta* series and 2.77×10^{10} and 4.90×10^8 s^{-1} for the *para* series. A related result has been reported for the tetranuclear Ir(III)/Ru(II) mixed-metal species **7** (although not linear, this species is briefly discussed here for convenience reasons) [214]. In **7**, both Ir-based emission ($\lambda = 572$ nm, $\tau = 2.9$ μs) and Ru-based emission ($\lambda = 682$ nm, $\tau = 82$ ns) are present, showing that photoinduced energy transfer from the

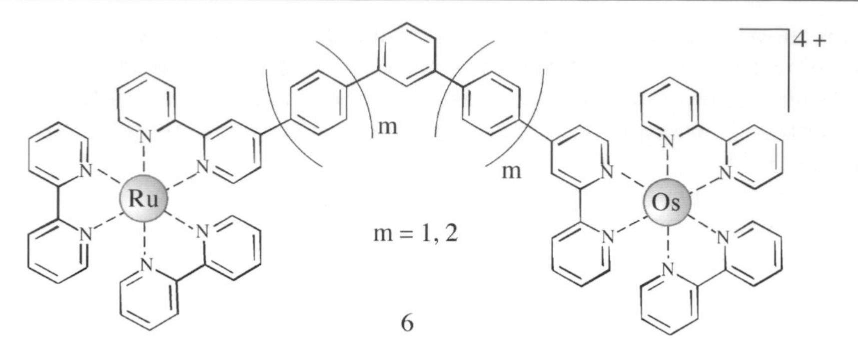

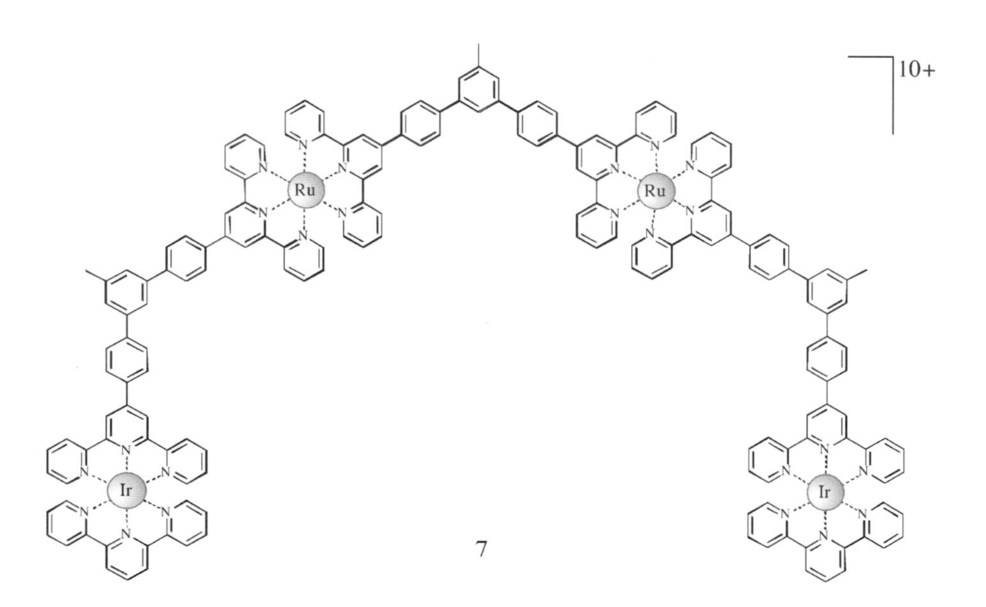

Ir(III) chromophores to the Ru(II) units is inefficient at room temperature in fluid solution. This suggests that the Ir-to-Ru photoinduced energy transfer rate constant in this tetranuclear species is lower than the intrinsic rate constant for Ir decay (about $3.7 \times 10^5\ s^{-1}$), in spite of the nonnegligible driving force (about 0.3 eV, from emission data). At 77 K, energy transfer from Ir-based to Ru-based chromophores is quantitative because of the much longer lifetime (205 μs) of the excited state of the Ir-based units. Indirectly, the room- and low-temperature results tend to suggest that Ir-to-Ru energy transfer in the tetranuclear mixed-metal species would occur with a rate constant of the order of $10^4\ s^{-1}$. The apparent discrepancy with the relatively fast energy transfer rate constant for the Ru–Os species with three phenylene unit bridges of the *meta* series discussed above (having a similar bridge to the Ir/Ru tetranuclear system here discussed) shows that the energy transfer

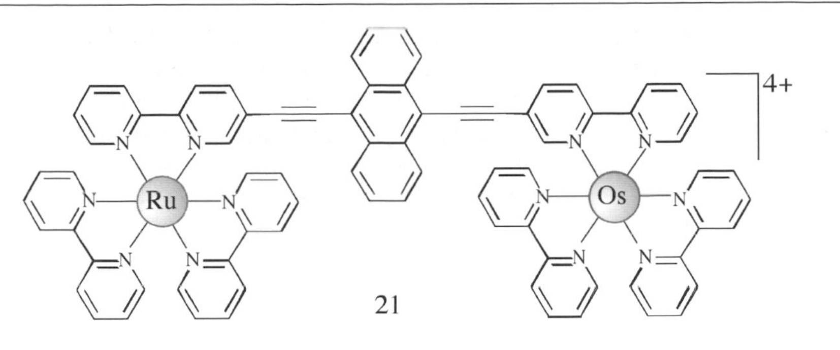

21

polyacetylenic backbone. For the phenyl-containing bridge system, the triplet state of the bridge is higher in energy than both the Ru(II) and Os(II) MLCT levels. Energy transfer takes place directly from the Ru(II) chromophore to the Os(II) one via a superexchange-assisted Dexter mechanism. In the naphthyl-bridged Ru–Os species, the triplet state of the bridge is intermediate in energy between donor and acceptor levels: the energy transfer from the Ru(II) subunit to the Os(II) one occurs in a stepwise manner, first to the central bis(alkyl)naphthalene unit of the bridge and then to the Os(II) site. In the anthryl-bridged species, the bis(alkyl)anthracene triplet is lower in energy than both Ru- and Os-based MLCT states, and the bridge plays the role of an energy trap [230].

The last Ru–Os compound discussed above has some similarity with the Ru–anthracene–Os species **22** [231, 232]. In this species, missing the ethynyl groups, the anthracene triplet lies in between the Ru donor and Os acceptor energy transfer subunits, so the behavior of the bridge is similar to that of the naphthyl-bridged species mentioned above. However, in air-equilibrated solution the energy transfer rate constant significantly decreases with increasing irradiation time. This effect is due to the formation of singlet oxygen by bimolecular energy transfer from the Os(II) excited state. The singlet oxygen reacts with the anthracene unit to give a peroxide species which cannot behave as the intermediate "station" for energy transfer, so that the overall process is significantly slowed down. This complex was called a "self-poisoning" species [231, 232].

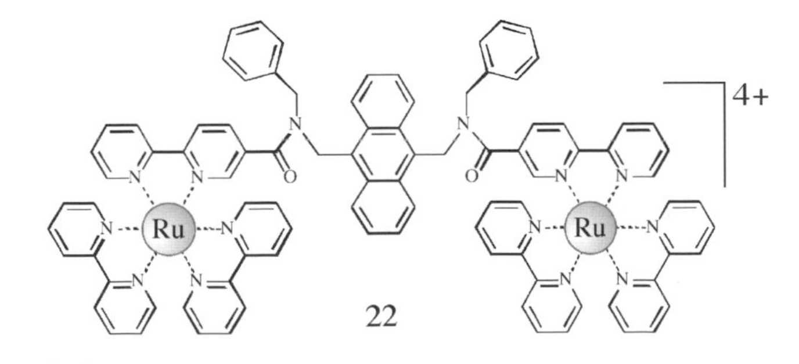

22

rates are significantly affected by the partner properties, as expected for both coulombic and (superexchange-assisted) Dexter-type mechanisms.

In the series of Ru–Os dyads with tridentate ligands **8** [215–217], both the donor Ru-based and acceptor Os-based MLCT states taking part in the energy transfer process involve the bridging ligand, while in the analogous, cyclometallated species **9** [215, 218] the donor and acceptor MLCT states are based on the peripheral ligands. In the larger systems, with two phenylene spacers, k_{en} for energy transfer from the Ru-based chromophore to the Os-based one is $5 \times 10^{10}\ s^{-1}$ for the non-cyclometallated species [215–217] and $< 2 \times 10^{7}\ s^{-1}$ for the cyclometallated species [215, 218]. These different results are mainly attributed to the fact that the energy transfer pathway is longer for the cyclometallated system, although it is suggested that the different nature of the bridge could also play a role.

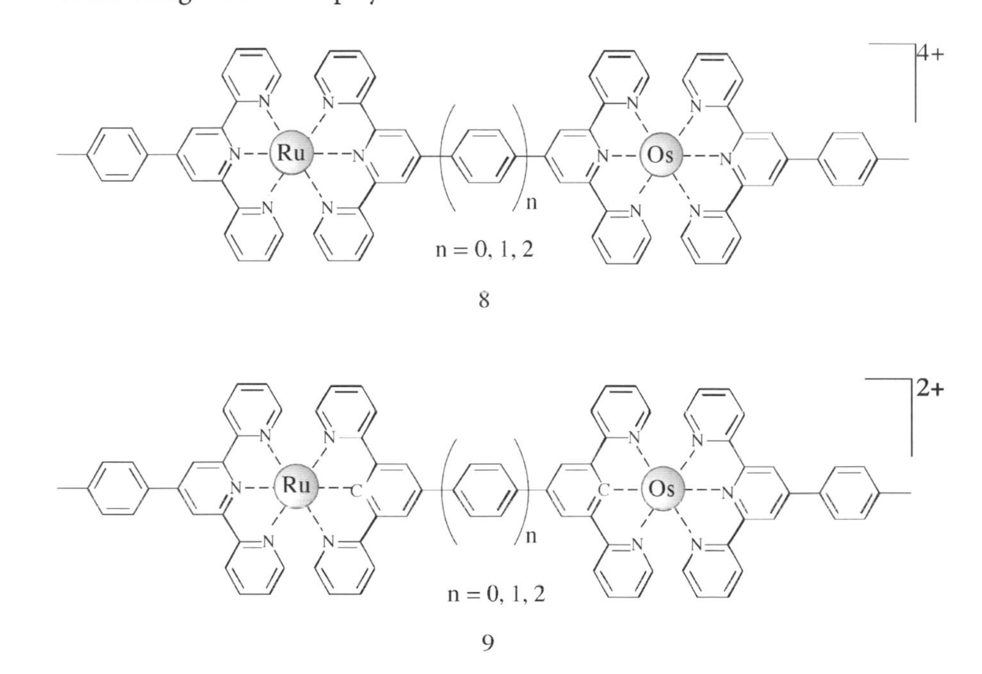

8

9

Possible effects of excited-state localization on intramolecular energy transfer kinetics are also shown by the results obtained for the two isomeric dinuclear Ru(II) species **10** and **11** [219]. In both complexes, energy transfer from the non-cyclometallated Ru subunit to the cyclometallated Ru subunit takes place by a Dexter mechanism. Ultrafast spectroscopic measurements yield different energy transfer time constants for the two isomers, with that related to the bridge-cyclometallated complex (2.7 ps) being faster than that related to the terminal-cyclometallated one (8.0 ps). This difference is explained in terms of different electronic factors for Dexter energy transfer. The lowest MLCT excited state in the Ru cyclometallated unit of the dinuclear

Another spacer subunit which allows for rigidity and controlled directionality is the alkynyl group. Dinuclear species incorporating an unsaturated polyacetylenic backbone in the spacer (**18**) have been extensively investigated [175, 226]. Energy transfer (electron exchange mechanism) from the Ru(II) chromophore to the Os(II) one takes place with a rate constant of 7.1×10^{10} and 5.0×10^{10} s^{-1} for $n = 1$ and $n = 2$, respectively [175]. The β attenuation factor [227] for the polyacetylenic systems was calculated to be 0.17 $Å^{-1}$, indicating that the "electron conduction" for energy transfer through alkyne bridges is more efficient than that through oligophenylenic spacers [228].

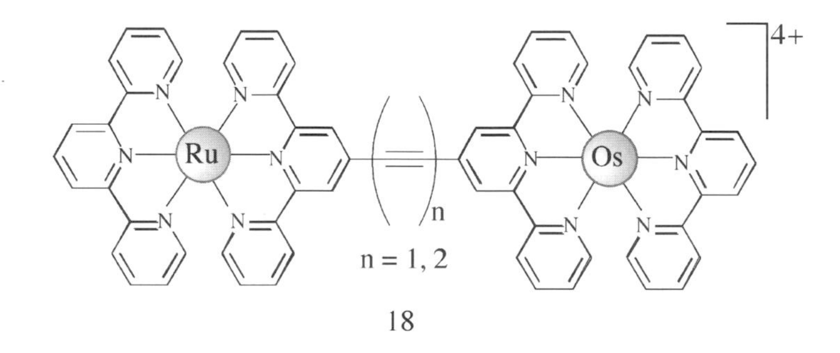

Several other differently connected multicomponent species based on Ru(II) chromophores as donors and incorporating polyacetylenic bridges have been studied. Interested readers can find information in [175, 176, 226, 229]. A special comment is warranted by the species **19–21** [230]. These compounds indicate the effect of the incorporation of additional units into the

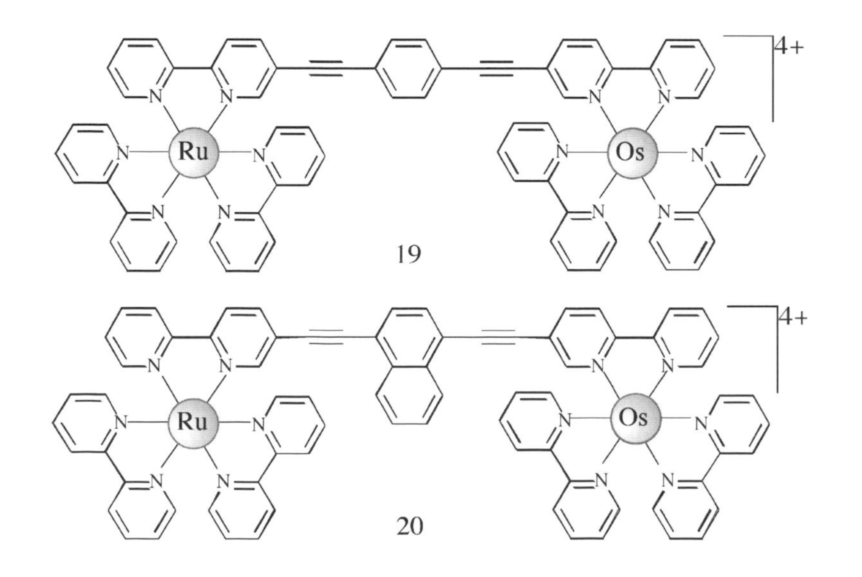

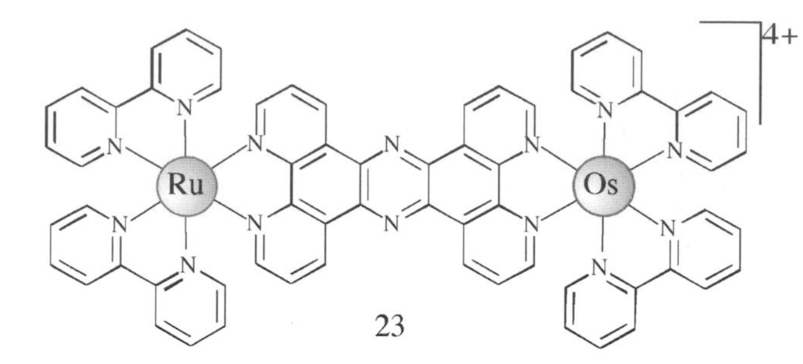

Another example showing an "active" role of the bridge in mediating intercomponent transfer processes involving Ru(II) species is evidenced by **23** [233]. In this species, there are two close-lying MLCT states per metal center involving the bridging ligand (leaving aside the MLCT state involving the peripheral ligands), because of the particular nature of the bridge (see Sect. 5.9). The higher energy of such MLCT states ($MLCT_1$) involves a bridging ligand orbital mainly centered in the bpy-like coordinating site (LUMO+1), and the lower energy one ($MLCT_0$) is localized on the central phenazine-like site (LUMO). Light excitation of the Ru-based chromophore populates the singlet $MLCT_1$ state, which rapidly decays to its triplet counterpart. Direct light excitation into the singlet $MLCT_0$ level (and successive population of its triplet) is inefficient because of the negligible oscillator strength of the transition. For Ru-to-Os energy transfer, two possible pathways are possible: (1) Ru-to-Os energy transfer at the $^3MLCT_1$ level (EnT), followed by $^3MLCT_1$-to-$^3MLCT_0$ relaxation within the Os(II) chromophore (a sort of intraligand electron transfer, ILET, within the Os(II) subunit); and (2) $^3MLCT_1$-to-$^3MLCT_0$ relaxation within the Ru(II) chromophore (ILET in Ru(II) subunit), followed by Ru-to-Os energy transfer at the $^3MLCT_0$ level. The situation is schematized in Fig. 13 [210, 233].

Interestingly, ultrafast spectroscopy shows that pathway 1 is followed in dichloromethane and pathway 2 prevails in the more polar acetonitrile solvent.

Oligophenyl bridges are reported to play "active" roles in the dinuclear Ir(III)–Ru(II) species **24–27** [234]. In this series of complexes, the Ru-based component is the energy transfer acceptor subunit. Indeed, Ru-based emission takes place in all the species at about 625 nm (lifetime about 200 ns) in aerated acetonitrile at room temperature and at about 590 nm (lifetime about 6 μs) in butyronitrile at 77 K, whereas the high-energy Ir-based chromophore has a very short excited-state lifetime, determined by time-resolved emission and subpicosecond transient absorption spectroscopy, slightly dependent on the bridge. The energy transfer rate constant is very weakly slowed down by increasing the bridge length, passing from 8.3×10^{11} s^{-1} for the species with two phenyls as spacer to 3.3×10^{11} s^{-1} for the species with five interposed phenyls. The apparent attenuation parameter β for energy transfer rate con-

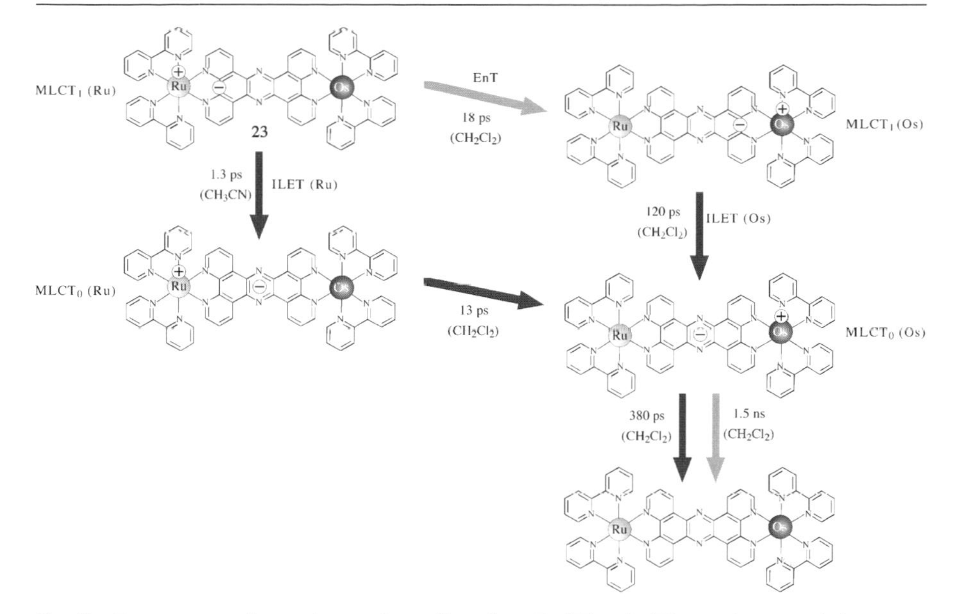

Fig. 13 Energy transfer pathways in a dinuclear Ru(II) – Os(II) species containing a π-extended bridge

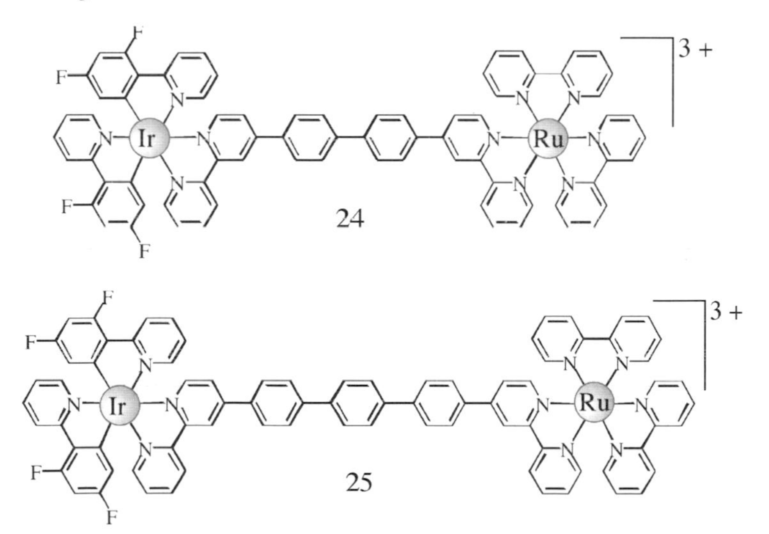

stant would be 0.07 Å^{-1}. However, increasing the bridge length changes the excited-state energy level of the bridge itself, and it is proposed that the MLCT excited state of the Ir center assumes an increasing LC character involving the bridge orbitals as the number of phenyls increases. As a consequence, metal–metal separation does not reflect the effective donor–acceptor separation for the energy transfer process; with the donor excited state largely involving the oligophenyl spacer, the energy transfer takes place by an in-

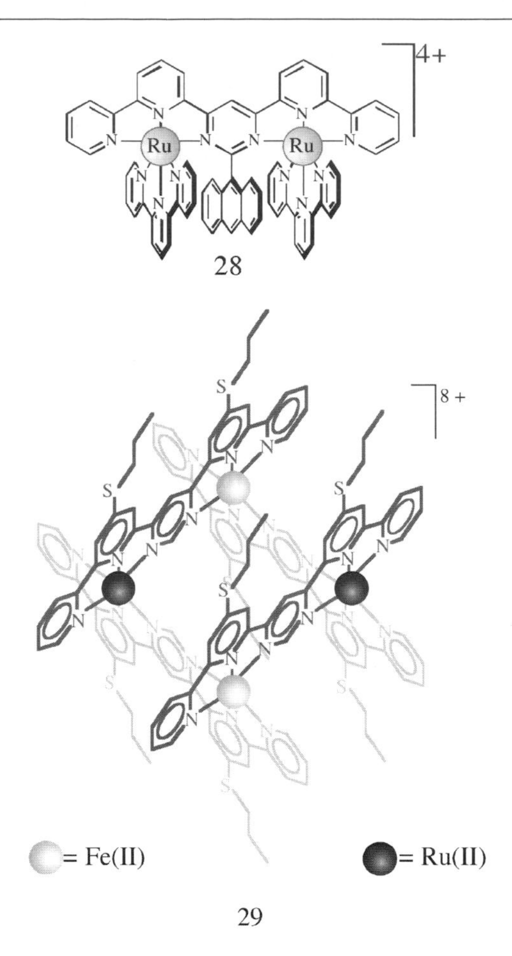

Ru(II) racks based on different molecular strands (**30**, **31**) have also been recently studied [240] (S Campagna, unpublished results). The pyrimidine-containing Ru complex exhibits a ^{3}MLCT emission with maximum at 758 nm

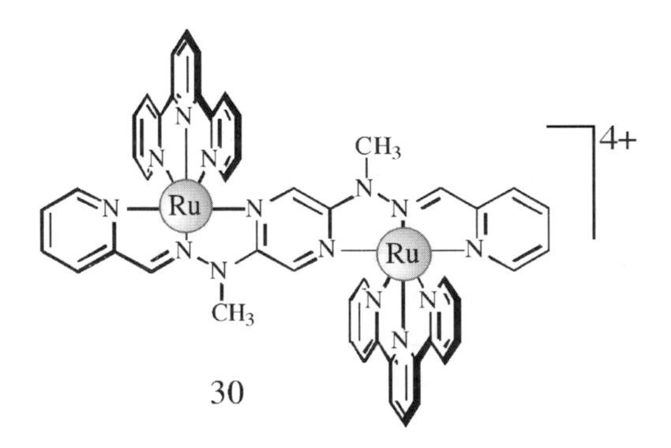

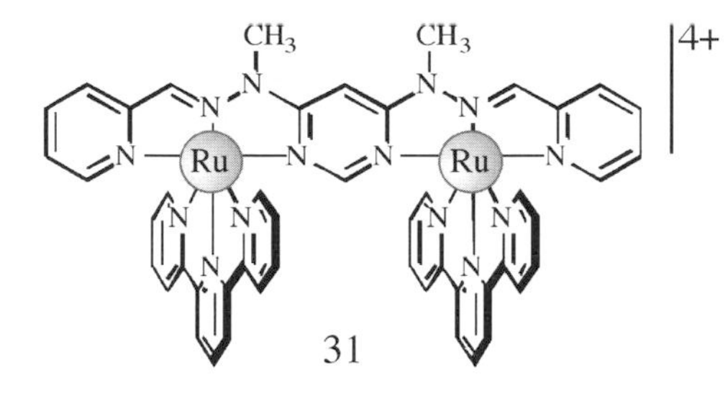

in acetonitrile at room temperature ($\tau = 30$ ns), which moves to 740 nm in nitrile glass at 77 K ($\tau = 335$ ns) [240]. The pyrazine-containing species exhibits very similar emission properties (room temperature, acetonitrile: $\lambda_{max} = 765$ nm; $\tau = 60$ ns; 77 K, nitrile glass: 750 nm; $\tau = 400$ ns) (S Campagna, unpublished results). For these species, the lowest (emitting) MLCT state(s) involve(s) the bridging ligands, as for the former rack-type complex discussed above.

5.2.2 Dendrimers

Luminescent Ru(II) dendrimers have been deeply investigated and the field has been reviewed recently [2, 241–245]. We will only mention some examples.

5.2.2.1 Dendrimers Containing Only One Metal Center Unit

The ruthenium compound **32** is a classical example of a dendrimer containing a luminescent ruthenium complex core surrounded by organic wedges. In this dendrimer, the 2,2′-bipyridine (bpy) ligands of the $\{Ru(bpy)_3\}^{2+}$-type core carry branches containing 1,3-dimethoxybenzene- and 2-naphthyl-type chromophoric units [246]. All three types of chromophoric groups present in the dendrimer, namely, $\{Ru(bpy)_3\}^{2+}$, dimethoxybenzene, and naphthalene, are potentially luminescent species. In **32**, however, the fluorescence of the dimethoxybenzene- and naphthyl-type units is almost completely quenched in acetonitrile solution, with concomitant sensitization of the $\{Ru(bpy)_3\}^{2+}$ core luminescence. These results show that very efficient energy transfer processes take place, converting the very short-lived (nanosecond timescale) UV fluorescence of the aromatic units of the wedges to the relatively long-lived (microsecond timescale) orange luminescence of the metal-based dendritic core. This dendrimer is therefore an excellent example of a light-harvesting antenna system as well as of a species capable of acting as a frequency converter. It should also be noted that in aerated solution the phosphorescence intensity of the dendritic core is more than twice as intense as that of the

based on the 2,3-dpp bridging ligand is fast and efficient, direct downhill energy transfer between partners separated by high-energy subunits is much slower and can be highly inefficient. This problem has been overcome (1) by using a third type of metal center, namely a Pt(II) one, to prepare decanuclear species (second-generation dendrimers) having different metal centers in each "generation" layer (schematically, $OsRu_3Pt_6$ species) [263] or, more recently, (2) in a heptanuclear dendron where the barrier made of high-energy subunits is bypassed via the occurrence of consecutive electron transfer steps [264]. Quite interestingly, this latter study suggests that long-range photoinduced electron transfer processes do not appear to be dramatically slowed down by interposed high-energy subunits in this class of dendrimers.

The efficiency of energy migration in 2,3-dpp-based dendrimers has attracted a large interest for the potential use of these species as synthetic antennae in artificial photosynthesis processes, and this has stimulated detailed kinetic investigations by means of ultrafast techniques. Studies on dinuclear model compounds have shown that esoergonic and isoergonic energy transfer between nearby units occurs within 200 fs, probably from nonthermalized excited states [168]. A direct consequence of such results is that energy transfer involving singlet states can compete with intersystem crossing. This conclusion is supported by the fact that the energy transfer from the peripheral Ru(II) subunits to the central Os(II) core in a tetranuclear $OsRu_3$ dendrimer takes place both by a singlet–singlet pathway, with a lifetime of less than 60 fs, and by triplet–triplet energy transfer, with a lifetime of 600 fs [169, 170]. The finding of singlet–singlet energy transfer is a particularly important result, since it indicates that the idea that any excited-state process involving metal polypyridine complexes had to be ascribed only to triplet states should be taken with caution when a significant electronic coupling between donor and acceptor is present. In some way, this finding also parallels the results obtained for photoinduced injection of electrons into semiconductors [167, 265–271].

An extension of this kind of antenna is a first-generation heterometallic dendrimer with appended organic chromophores like pyrenyl units (**36**) [272]. In this species, consisting of an Os(II)-based core surrounded by three Ru(II)-based moieties and six pyrenyl units in the periphery, 100% efficient energy transfer to the Os(II) core is observed, regardless of the light absorbing unit. A detailed investigation of the excited-state dynamics occurring in this multicomponent species on exciting in the UV region (267 nm) has also been performed [273]. Transient absorption spectra (in the range 420–700 nm) for the various intermediates have been reported by the acquisition of evolution-associated difference spectra.

Energy transfer processes from the nonrelaxed and relaxed S_1 state of the peripheral pyrenyl chromophores to the lowest-lying Os-based MLCT triplet excited state occur with lifetimes of about 6 and 45 ps, respectively [273]. Subpicosecond energy transfer from the excited Ru manifold to the Os-based

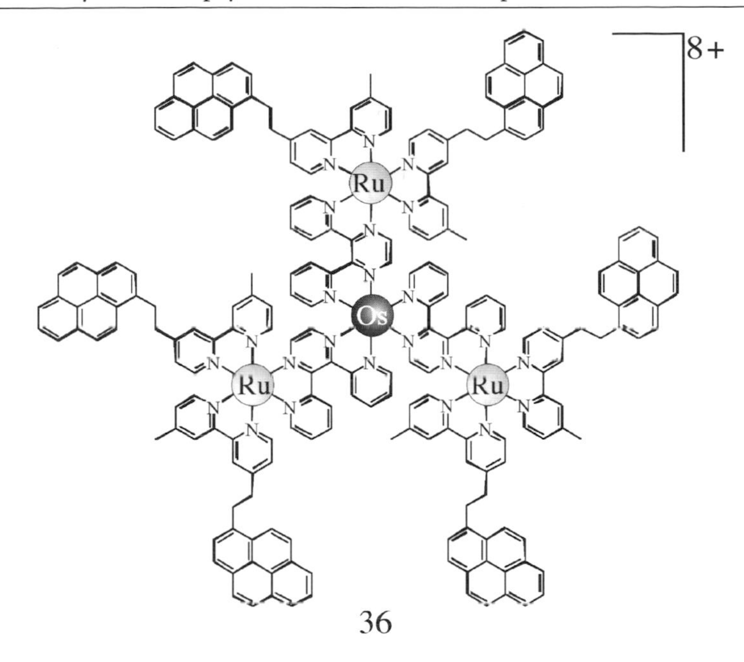

36

chromophore and interconversion between the initially prepared S_3 state and the low-lying S_1 level within the pyrenyl subunits have also been evidenced. The rate constant of the energy transfer from the pyrenyl groups to the Ru/Os excited state manifold is in good agreement with the Förster mechanism when the relaxed S_1 pyrene state is taken into account. Energy transfer from the nonrelaxed state most likely involves folded conformations in which the pyrenyl subunits are strongly interacting with inner subunits of the tetranuclear core. Such interactions were also suggested by the ground-state absorption spectrum of the compound [272].

Because octahedral metal complexes can exist in two chiral forms, Λ and Δ, it could be expected that the photophysical properties of dendrimers containing metal complexes as branching centers could be different for the various isomers (the situation can be even more complicated in the case of geometrical isomers). However, the investigation of optically pure isomers of dinuclear and dendritic-shaped tetranuclear species (**37** is the general structural formula of the tetranuclear systems: optical geometry is not evidenced) has shown that stereochemical isomerism does not cause any sizeable difference, at least for the studied compounds [194].

In **37** the emissive state, which involves the peripheral Ru(II) centers, does not have a sizeable absorption counterpart, since it is a special type of charge-separated state, with the formal "hole" localized on a peripheral Ru(II) center and the "electron" localized on an orbital mainly centered on the pyrazine moiety of the bridging ligand: the absorption related to such a state has negligible oscillator strength, due to the poor overlap of the orbitals involved

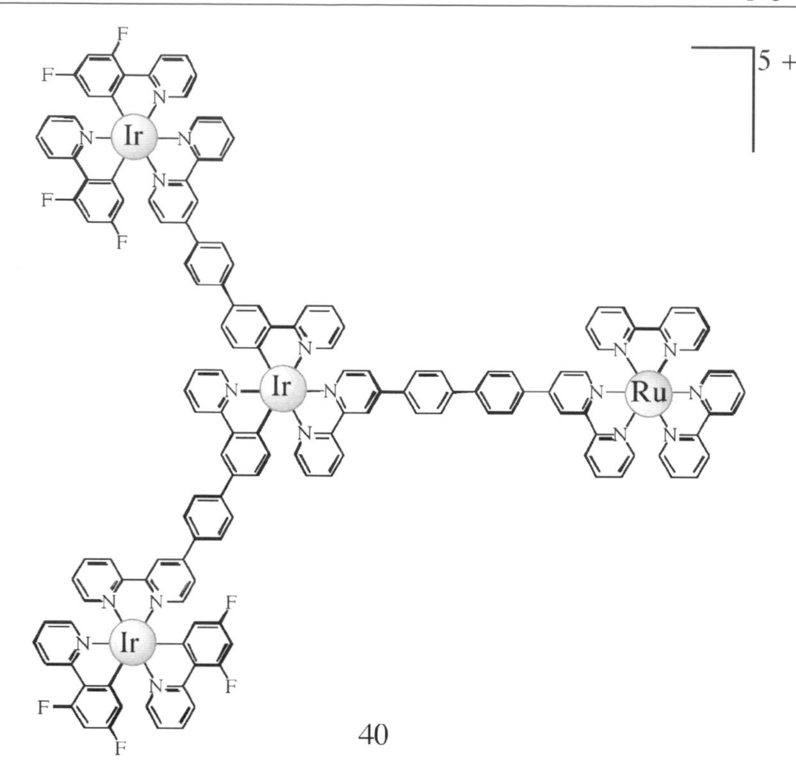

efficient, which confirms that in this type of artificial antenna metallodendrimer, long-range electron transfer can be quite effective, as in the case of the heptanuclear complex mentioned above [264], and could suggest interesting options to build up integrated donor–antenna–acceptor systems. This discussion leads us directly to the next section.

5.3 Donor–Chromophore–Acceptor Triads

Triad systems (Fig. 14) are key components of the early events in artificial photosynthesis: the light energy collected by the chromophore (P) is transformed into chemical (redox) energy by a sequence of electron transfer steps involving electron donor (D) and electron acceptor (A) units, ultimately leading to charge separation [208, 209, 228]. Charge separation is probably the most important photoinduced process on Earth, so it is not surprising that many triads based on Ru(II) complexes have been prepared and studied in the last 20 years [228, 280]. It should be noted that there are literally dozens of dyads based on Ru polypyridine complexes [228, 281]. Only some examples of triads (that is, species where Ru(II) chromophores are simultaneously coupled to electron donor and acceptor units) are discussed here.

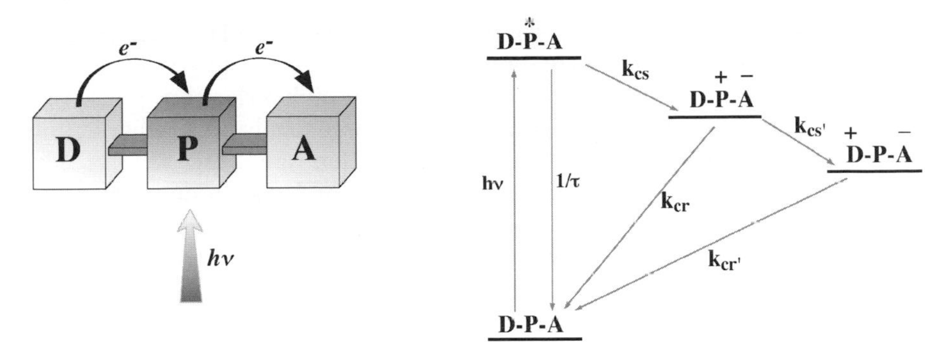

Fig. 14 Schematization of a triad for photoinduced charge separation. P, chromophore; D, donor; A, acceptor; cs, primary charge separation; cr, primary charge recombination; cs′, secondary charge separation; cr′, final charge recombination

Structurally speaking, Ru complexes with tridentate ligands like terpy are ideal systems for molecular triads: indeed, substitution at the 4 position of the central pyridine ring of terpy-like ligands allows a linear arrangement of subunits, with control of geometry and distance. Some of the first ruthenium triads based on such an arrangement are shown in Fig. 15 [282]. Whereas fully developed charge separation, with formation of the D^+-P-A^-

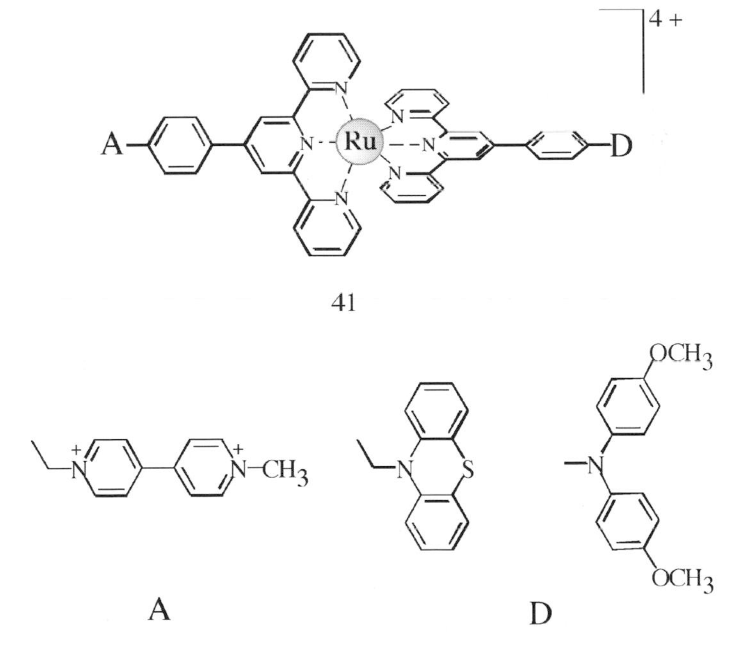

Fig. 15 Structural formulae of Ru(II) terpyridine triads containing acceptor (A) and donor (D) components

charge-separated state, did not take place in the triad with D = PTZ (PTZ = phenothiazine subunit), the formation of such a charge-separated species was inferred for the species with D = DPPA (DPPA = diphenylamino moiety) at 150 K, although it could not be evidenced spectroscopically because it did not accumulate as a consequence of a fast recombination rate [282].

In spite of less control of distance and orientation between donor and acceptor, better results have been reported for a series of systems exemplified by **42** [283–285]. The components of this series of triads are a tris-bipyridine Ru(II) chromophore covalently linked to one or two phenothiazine electron donors and to quaternized bipyridinium electron acceptors. The saturated alkyl chains bridging the molecular components are electrically insulating and flexible. This latter point is apparently a drawback since it does not allow for control of geometry. Moreover, even the octahedral arrangement of the bpy subunits adds some difficulties in defining the real structure: for example, geometrical isomers can also exist, since each bpy of **42** is non-symmetric. The compound **42** exhibited formation of the D^{+}–P–A^{-} charge-separated state in dichloromethane at room temperature with an initially reported efficiency of about 26%. Such a value was later corrected to about 86% by using a slightly different solvent (1,2-dichloroethane) [283, 286, 287]. Once formed, the charge-separated state decayed with a relatively fast rate ($k_{CR} = 6.3 \times 10^{6}\ s^{-1}$), corresponding to a lifetime of about 160 ns. On the basis of redox data, the charge-separated state stored about 1.3 eV.

Similar complexes were prepared that differed from one another by the length of the alkyl chains connecting the subunits and/or by changing

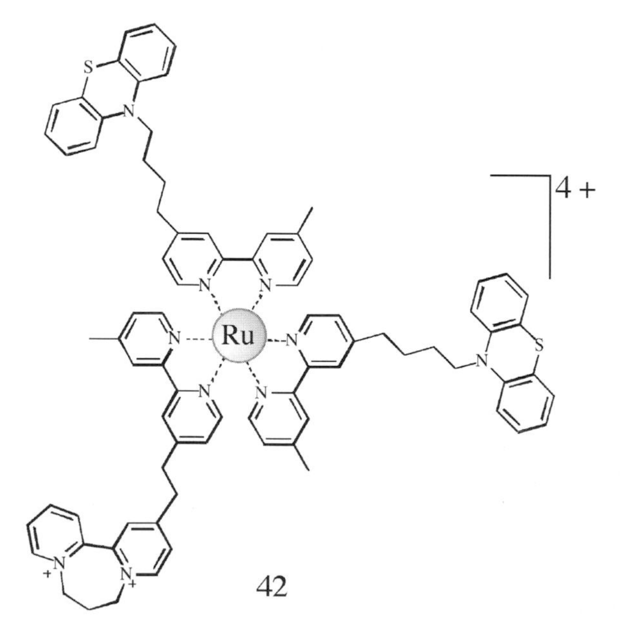

the methylene chains connecting the quaternary nitrogens (and, as a consequence, the reduction potential of the electron acceptor) [283–287]. In a homogeneous series of experiments performed on such species, however, the efficiency of charge separation does not change appreciably, remaining larger than 0.80, although the driving forces and the rate constants of the various electron transfer steps, as obtained by independent studies performed on isolated dyads of the type D–P or P–A, were different.

In D–P*–A systems, the fully developed charge-separated state can be obtained, in principle, by two different routes (excluding direct electron transfer from D to A): (1) a route initiated by oxidative quenching, that is, the series of events described by the sequence D–P*–A, $D–P^+–A^-$, $D^+–P–A^-$; and (2) a route initiated by reductive quenching, described by the sequence D–P*–A, $D^+–P^-–A$, $D^+–P–A^-$. Both routes can also take place simultaneously. The comparison between the photophysical properties of the various triads and the corresponding isolated dyads of this family of compounds indicated that the emission decay rates of any D–P–A triad never differed by more than a factor of two from those of the P–A dyads, although the absolute decay rate values changed by over a factor of 10^3 (over the whole collection of compounds). This prompted the authors to attribute the initial quenching event in all the D–P–A triads of this family to oxidative electron transfer, with formation of the $D–P^+–A^-$ intermediate, with the route initiated by reductive quenching playing a negligible role [288]. However, in all the P–A dyad systems, it was always impossible to detect the A^- radical anion [284], indicating that back electron transfer in the P–A dyads was faster than the forward, oxidative electron transfer. This posed some problems in justifying the efficiency of formation of the fully developed charge-separated state, where apparently reduction of P^+ from D in $D–P^+–A^-$ species efficiently competes with back electron transfer in the intermediate. In fact, this looks somewhat puzzling because the reductive electron transfer in D–P* dyads is reported to be of the order of 10^6 s^{-1} [289], while oxidative electron transfer in P*–A dyads ranges from 10^{10} to 10^7 s^{-1} [284, 285, 290–293] and, based on the circumstances mentioned above, back electron transfer in $P^+–A^-$ (and for extension in $D–P^+–A^-$), opposing the formation of the fully developed charge-separated state, could be even faster. To justify the experimental data, electron transfer from D to P^+ in $D–P^+–A^-$ should be about 1×10^{10} s^{-1} or faster. Therefore, the exceptional properties of these compounds as far as the efficiency of charge separation is concerned remained largely unexplained.

A recent paper has shed light on the photophysical behavior of these triads [288]. A series of new experiments, including transient absorption measurements, emission decay, and a careful examination of the ground-state absorption spectra of the triads and of various separated dyad components, suggested that in the D–P–A triads of this family an association between the tethered phenothiazine electron donor subunit and the Ru(II) chromophore takes place, in a folded conformation. The association is already present in the

parent first-order rate constant of $6.6 \times 10^3\ s^{-1}$. This species is produced with a low quantum yield (0.5%), stores about 1.3 eV, and recombines to the ground state with a lifetime exceeding 2 ms. In spite of the low efficiency of the charge-separation process, there are several interesting points: (1) in the absence of myoglobin, charge separation is not obtained at all; (2) the complete process is pH dependent; (3) the final charge-separated state lifetime is comparable with that of natural photosynthetic reaction centers; and (4) back electron transfer is regulated by protonation/deprotonation of distal histidine moieties, which appears to be needed to reduce the Mb(Fe^{IV} =O) subunit. The low efficiency of the overall process is mainly attributed to charge recombination within the Mb($Fe^{III}OH_2$)$^+$-Ru^{2+}-BV^{3+} state, which efficiently competes with the deprotonation processes. This study highlights the potential of mixed synthetic–natural systems for obtaining long-lived charge separation.

5.4 Polyads Based on Oligoproline Assemblies

The interesting results obtained by organizing D–P–A triads on the structure of the amino acid lysine (Sect. 5.3) prompted the preparation of D–P–A systems assembled on oligoproline scaffolds, by means of solid-state peptide synthesis [294, 297–299]. One such system is **46**. In this species, a phenothiazine (PTZ) group acts as the electron donor and an anthraquinone (Anq) subunit plays the role of the electron acceptor. Oligoprolines were selected, since it is known that oligoproline chains of nine or more proline units fold into stable helices even with large functional groups on the proline units. The terminal segments allow the helix to begin and end with capped Pro_3 turns which prevent unwinding of the helix. For **46**, a fully developed charge-separated state is gained in acetonitrile solution with good efficiency (53%). The charge-separated state stores 1.65 eV relative to the ground state and returns to the ground state with a rate constant of $5.7 \times 10^6\ s^{-1}$ (τ = 175 ns) [299, 300]. Quenching of the Ru-based excited state is domi-

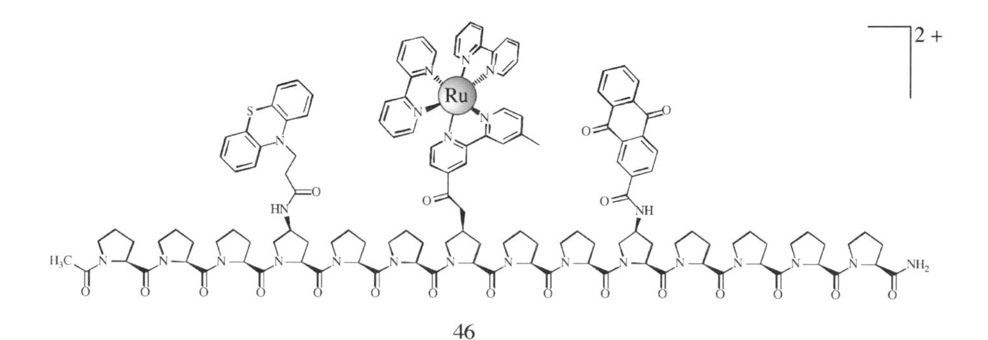

46

nated by reductive electron transfer involving the PTZ electron donor, in a process which is largely solvent dependent, as is the driving force of the process. Then, there is a fast electron transfer from the reduced metal chromophore to the Anq subunit, which yields the charge-separated state. There is a strong solvent-dependent competition between such a second electron transfer, which allows for fully developed charge separation and back electron transfer in the D^+–P^-–A intermediate. The consequence of this solvent dependence is that going from 1,2-dichloroethane to dimethylacetamide, the efficiency of charge separation changes from 33 to 96%. Also, the charge recombination is solvent dependent, and the electronic coupling between PTZ^+ and Anq^- was calulated to be about 0.13 cm^{-1} [300].

A more elaborated polyad based on the formerly described systems is the D–P–P–A tetrad **47** [301]. This is the evolution of a system quite related to **46**, where substituents on the terminal bpy ligands of the metal chromophore are used to favor the thermodynamics of the (reductive) first electron transfer step. This modification led to an efficiency of charge separation of 90% in acetonitrile for the corresponding triad. In the tetrad, 13 proline spacers are present between PTZ and Anq. The efficiency of formation of the charge-separated state in the tetrad is 60% and its lifetime is 2 μs ($k_{CR} = 5.0 \times 10^5\ s^{-1}$). Excitation can occur in both the Ru chromophores, but apparently the result is not identical. Excitation of the Ru(II) complex adjacent to the PTZ electron donor gives the D^+–P^-–P–A system. To produce the fully developed D^+–P–P–A^+ species, it is proposed that a stepwise mechanism occurs, with the species D^+–P–P^-–A as an intermediate. Efficient, isoergonic electron transfer between the two chromophores is therefore foreseen. Excitation of the Ru(II) complex adjacent to the electron acceptor Anq subunit would be unproductive in a direct sense, since oxidative electron transfer by Anq is unfavorable thermodynamically and direct quenching from the PTZ unit is unlikely because of the large distance. However, even excitation of this Ru(II) chromophore can become productive, provided that isoergonic energy transfer to the Ru(II) chromophore adjacent to the PTZ unit takes place. Since the quantum efficiency of formation of the charge-separated state is 60%, and considering that excitation of the two identical Ru(II) chromophores

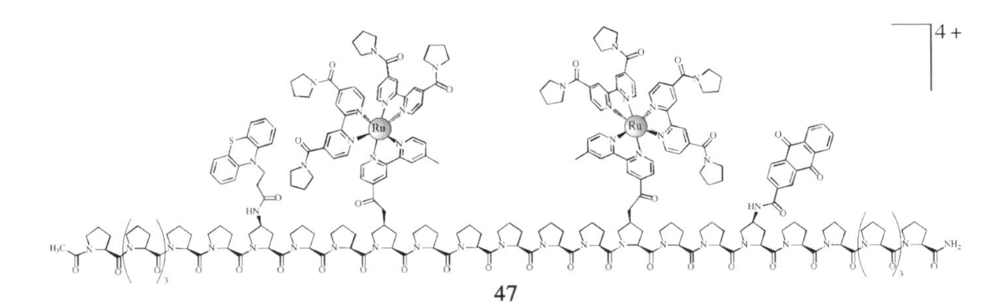

47

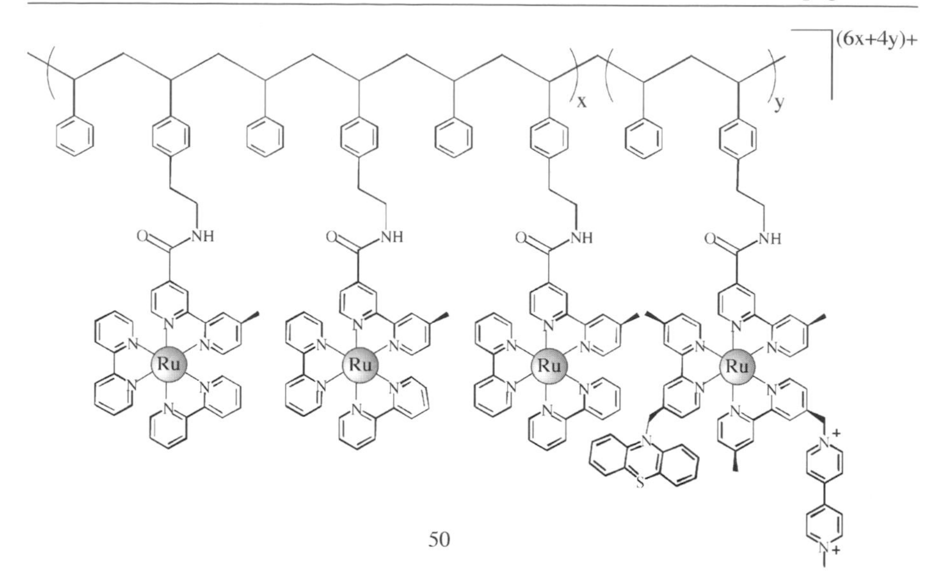

containing 17 normal Ru chromophores and three RCs, the D^+–A^+ charge-separated state is formed, as shown by transient absorption spectroscopy. Emission at the Ru(II) chromophores was quenched by about 34% compared to the homopolymer containing 20 "normal" Ru chromophores. Energy transfer from the normal Ru chromophores to the RC sites was favored by − 0.1 eV. It was shown that about 50% of the charge-separated state was formed during the 7-ns laser pulse, indicating intrastrand sensitization, with the charge-separated state formed by direct excitation at the RC complex and by excitation to nonadjacent Ru chromophores followed by energy migration to the RC sites. In this system, 1.15 eV is stored in the charge-separated state and the efficiency of the process varies from 12 to 18% depending on laser irradiance, indicating excited-state annihilation at high irradiance. Charge recombination is similar to that of the "isolated" RC, but an additional long-lived transient (formed in low efficiency, about 0.5%) was observed, which decayed by second-order kinetics with $k = 48\ M^{-1}\ s^{-1}$. This long-lived transient was attributed to polymers in which D^+ and A^- were formed on different RC units, by invoking mechanisms in which electron transfer quenching by oxidative or reductive electron transfer in a RC site is followed by intrastrand hole or electron transfer to a second RC site [311].

5.6 Photoinduced Collection of Electrons into a Single Site of a Metal Complex

An essential property of natural photosynthesis is the collection of multiredox equivalents at specific sites. Indeed, all the important light energy storage processes require more than one electron to operate: for example, reduction of H^+ to H_2 is bielectronic, and oxidation of oxygen in water to produce O_2 is a four-electron process. Reduction of CO_2 to the high-energy content glucose species is also a multielectron process. Whereas artificial systems capable of performing photoinduced charge separation have been reported, species able to collect, by successive photoinduced processes, more than one single electron (or hole) in *one specific site* of their structure are very rare. These species differ from polymers or dendritic species, which are also able to reversibly store more than one single electron (or hole) in their structure (in several, roughly identical, but *spatially separated sites*), since the accumulated charges should be located in a single subunit and, at least in principle, could be more easily delivered simultaneously to a unique substrate.

A breakthrough in this field was the study of the two dinuclear Ru complexes **51** and **52** [312]. These complexes are indeed able to collect two electrons (and two protons) and four electrons (and four protons), respectively, within their bridging ligand moieties upon successive light excitation and in the presence of sacrificial donor species. In a typical (schematized) sequence of events involving **51**: (1) light excitation produces a MLCT state involving the bpy-like subunit of the bridge; (2) a charge shift takes place from the bpy-like bridge moiety to the inner, phenazine-like portion of the bridge, so producing a sort of charge-separated state; (3) the sacrificial donor, a triethylamino (TEA) species, reduces the Ru(III) center, so restoring the chromophore; (4) the reduced central moiety of the bridge adds a proton (originated from irreversible TEA oxidation), so reaching charge neutrality; and (5) the sequence of events 1–4 is repeated and two electrons and two protons are collected [312]. However, a recent refinement of the ultrafast spectroscopic results has evidenced that the product of step 2, initially identified as a sort of charge-separated state [313], receives a significant contribution also from a bridge-centered triplet state [314]. The overall process is perfectly reversible, and **51** is fully restored on leaving molecular oxygen reaching the complex [312]. For **52**, the formerly described sequence of events is repeated four times, thanks to the presence of the quinone subunits responsible for the addition of two extra electron/proton couples [312]. All the various steps of the multielectron processes occurring in **51** have also been characterized by UV/Vis spectroscopy and each intermediate has a unique

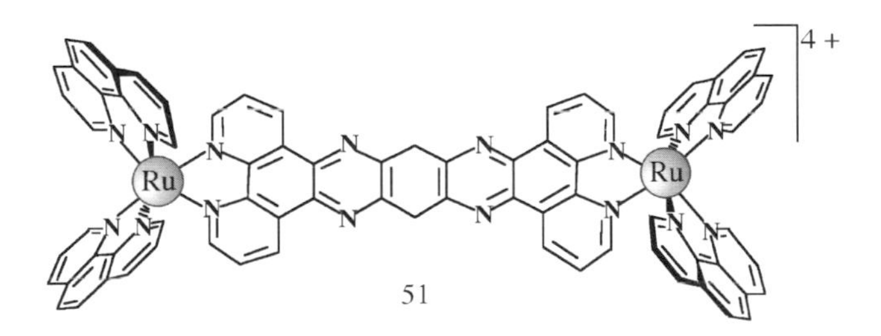

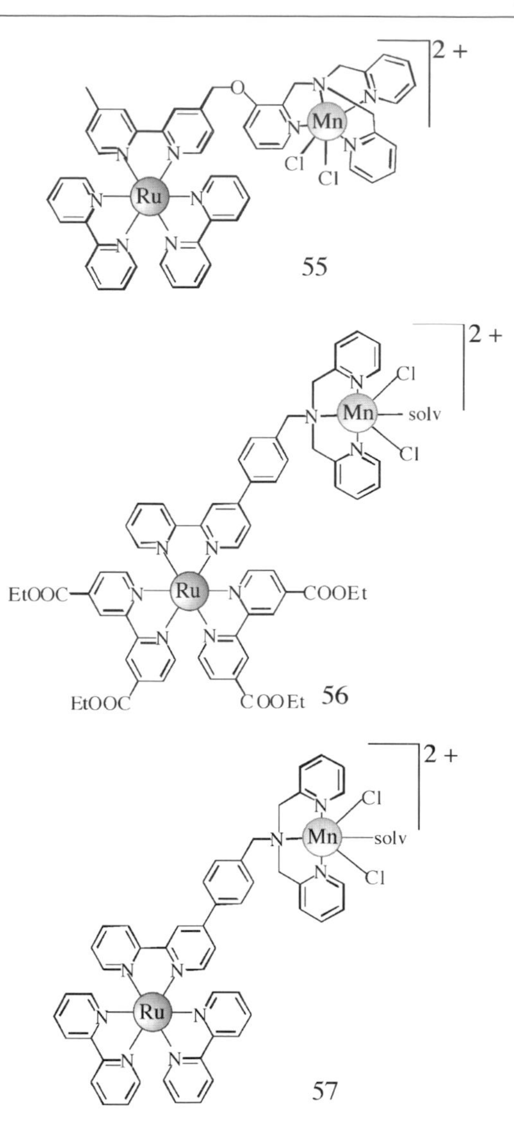

< 50 ns to 10 μs, depending on the complex [326]. This study demonstrated that the reorganization energy for the Mn(II)-to-Ru(III) electron transfer was quite large (1.4–2.0 eV), suggesting significant inner reorganization of the manganese moiety during the process [327]. As an obvious consequence, very fast Mn(II)-to-Ru(III) electron transfer could not be expected. A further complication was that the manganese subunit could directly quench the excited ruthenium chromophore by Dexter energy transfer, competing with electron transfer quenching by the sacrificial acceptor for short Ru–Mn distances. These arguments led to the preparation of more elaborated systems in which intermediate donor species were interposed between ruthenium and

manganese subunits [328–330], with phenolate and tyrosine moieties playing the role of an intermediate donor. In most cases, proton-coupled electron transfer processes took place.

To achieve multielectron catalysts, more than one manganese ion was included in the systems. Figure 16 shows some examples [329, 331, 332] of Ru(II) species covalently linked to Mn dimers or trimers via phenolate ligands. In particular, for the Ru–$Mn_2^{II,II}$ complex **58** reported in the figure, repeated flashes in the presence of a Co(III) sacrificial electron acceptor allowed three successive one-electron oxidations of the manganese moiety by the photooxidized Ru(III) subunit [333], as evidenced by the disappearance of the characteristic $Mn_2^{II,II}$ signals and the appearance of the characteristic $Mn_2^{III,IV}$ signals in EPR experiments. Manganese oxidation was suggested to involve a ligand exchange, in which acetate is released and water molecules are bound to form a di-μ-oxo bridge (see the reaction scheme in Fig. 16). According to the authors, this was the first example of a light-driven, multiple oxidation of a manganese complex attached to a photosensitizer. The ligand exchange at the manganese sites (presumably occurring in the $Mn_2^{III,III}$ state, that is, after second electron release from the initial $Mn_2^{II,II}$ center) is functional to the overall process, as it allows introduction of negative

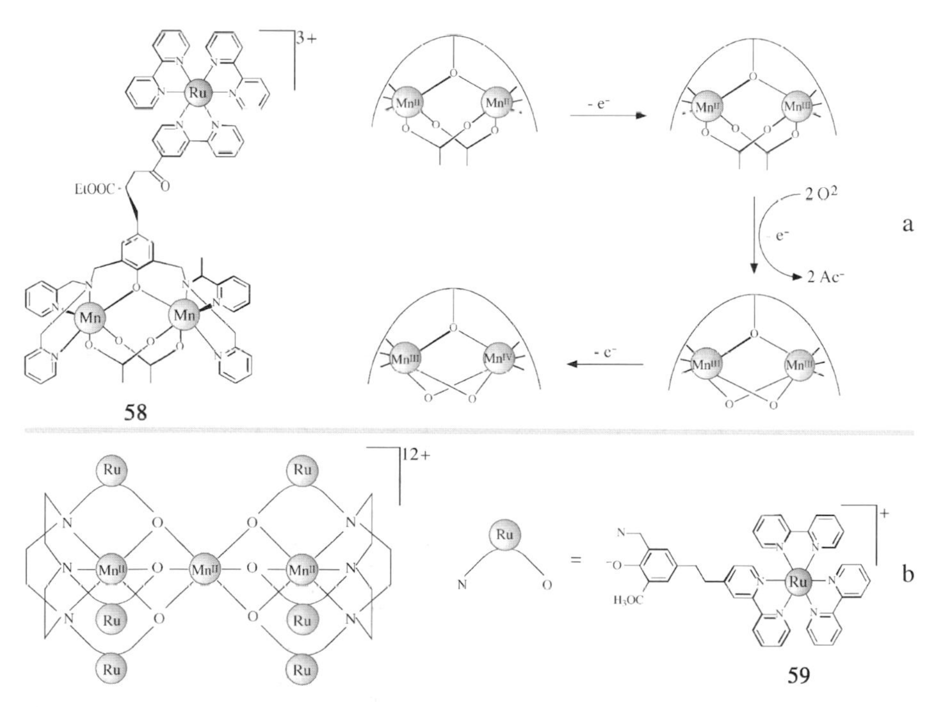

Fig. 16 Electron transfer from the manganese moiety to the photooxidized Ru(III) in **a** a Ru – $Mn_2^{II,II}$ complex and **b** a Ru – $Mn_3^{II,II,II}$ complex

tained in the dark or when one of the key components (the Ru(II) chromophore, the Pd catalyst, or the particular bridging ligand) is missing. Interestingly, the same compound where the bridging ligand between the two metal sites is a bipyrimidine unit does not evolve molecular hydrogen. It is also reported that the amount of photocatalytically formed hydrogen depends strongly on the TEA concentration and the exposure time, and chloride ions inhibit the reaction. The amount of hydrogen produced increases steadily and levels off after 1200 min. After about 1800 min no more hydrogen is produced. The rate of hydrogen formation increases with increasing TEA concentration for low TEA concentration, but becomes independent at a TEA concentration $> 0.86\ \mathrm{mol\,L^{-1}}$, where it is about $1600\ \mathrm{nmol\,min^{-1}}$. Although detailed mechanistic data are not available, the authors suggest as a first step of the process a twofold photoinduced reduction of the compound by TEA, analogously to what was reported for the related photoinduced electron collection system **51**. Probably reduction is concomitant with proton extraction from TEA oxidation products: TEA should therefore be the proton source. The successive step should be reduction of the protons at the nearby Pd center. This latter step probably passes through a temporary chloride loss. The same paper also reports the photocatalyzed selective hydrogenation of tolane to *cis*-stilbene, accomplished by the same compound. Analogously to **60**, the Ru(II) chromophore of **61** acts as the light harvester/photosensitizer (which also contains in its structure the electron acceptor subunit, that is, the phenazine moiety of the bridging ligand), while the role of the catalytic unit is here played by the Pd(II) center.

Molecular hydrogen evolution under visible light irradiation has also been reported for trimetallic species like **53** [348]. Mechanistic details are not available.

5.8.2 Other Photocatalytic Systems

The catalytic potential of heterometallic species containing Ru(II) and Re(I) chromophores for the conversion of CO_2 to CO has been recently shown [349]. Compound **62** is one of the species in this regard. This investigation highlighted the fact that the photocatalytic activity is deeply influenced by the nature of both the bridging ligand and the peripheral ligands at the light-harvesting Ru(II) chromophore. The proposed mechanism is that upon light irradiation ($\lambda > 480$ nm) in DMF/triethanolamine (TEOA, acting as base) with 1-benzyl-1,4-dihydronicotinamide (BNAH) as sacrificial donor, the initially produced Ru-based MLCT state is reduced by BNAH. Then intrabridging ligand electron transfer occurs, with formation of the reduced rhenium subunit. This latter species is known to react with CO_2 upon Cl ligand loss [350, 351]. The reduction of CO_2 is bielectronic, so it is assumed that the second electron transfer follows a similar route. The most efficient species of this series of compounds, which is exactly **62**, exhibits a turnover number of 170.

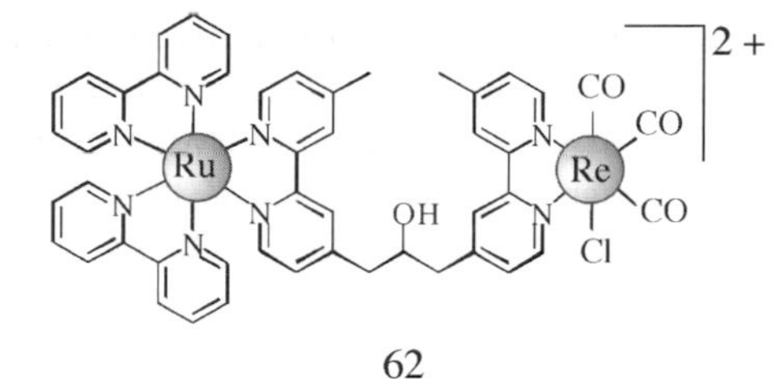

62

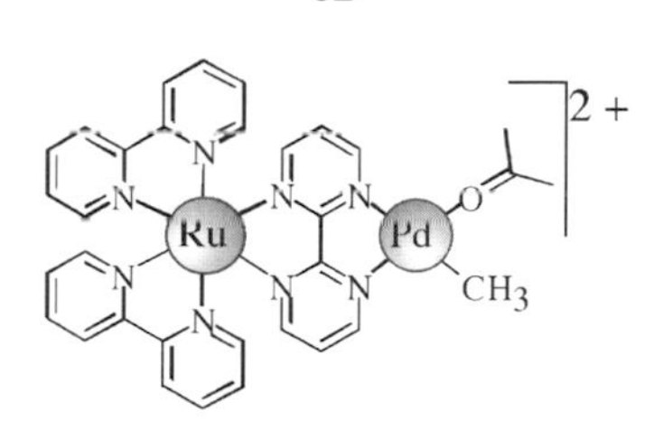

63

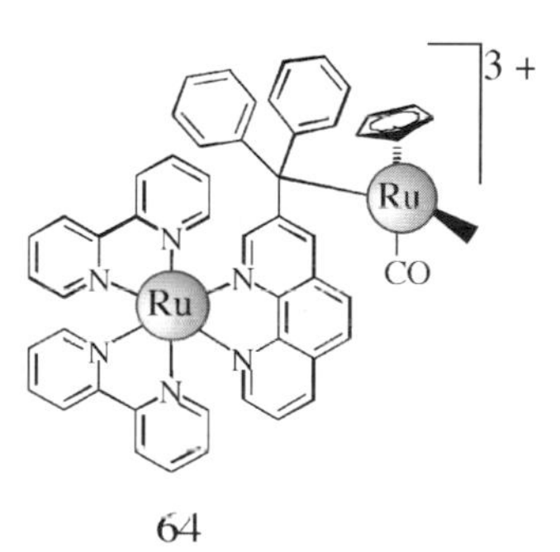

64

Two other dinuclear species (**63** and **64**) have been reported to show photocatalytic activity for specific reactions. The Ru–Pd dimer **63** is active for the photocatalytic dimerization of 1-methylstyrene [352], and the turnover numbers of the photocatalyzed reaction (> 90 within 4 h) and the high selectivity compete well with thermal catalytic systems. Compound **64** is active for the conversion of *trans*-4-cyanostilbene to its *cis* form [353]. Other aspects of the last mentioned works and of similar systems are also commented on in a very recent paper [354]. Photocatalytic processes based on photoelectrochemical cells in which the Ru chromophores are physically interfaced to electrodes or other solid systems are reported later.

5.9 Photoactive Molecular Machines Able to Perform Nuclear Motions

In the last 10 years there has been great interest in designing molecular machines [280]. As machines of the macroscopic world, even molecular machines need energy to operate, and a suitable form of energy to power

tization, in aqueous solution the dominant (lowest energy) excited state is the MLCT involving the phenazine-like dppz subunit (populated by a sort of "charge shift" decay from the directly excited MLCT level), which deactivates largely by nonradiative processes. When the complex interacts with DNA, the MLCT state involving the bpy-like dppz subunit becomes dominant, and since such a state has better luminescent properties, the luminescence of the complex is switched on. Looking in more detail, the interplay among the various states, in this and related species and in the absence and presence of DNA, is more complicated, as demonstrated by various theoretical [379] and experimental techniques, including transient absorption femtosecond spectroscopy [370, 371, 378, 380, 381] and time-resolved resonance Raman spectroscopy [372, 373, 382]. Details can be found in the original references.

Besides **67**, many other Ru(II) polypyridine complexes have been reported to exhibit luminescence enhancement in the presence of DNA. In most cases, the luminescence enhancement is moderate and can be assigned to the protection offered toward oxygen quenching by DNA structures to the surface-attached Ru(II) complex (which can bind, essentially for electrostatic reasons, to the major or minor grooves). If the interaction is limited to surface binding, the luminescence enhancement is usually within 20–40% in the presence of oxygen, whereas it is negligible in deoxygenated samples. Several compounds, however, exhibit noticeable luminescence enhancement (one order of magnitude or higher): in most of these cases, the compounds quite often have a ligand with a large, flat framework and intercalation takes place. The compound **68** [383], which is nonemissive in water solution and strongly emissive in organic solvents ($\lambda = 610$ nm; $\tau = 1.1$ µs; $\Phi = 0.12$) is an exception. In water, the presence of DNA switches the luminescence on. The authors suggest that in water solvent-specific interactions with the amido moiety promote radiationless decay of the (potentially) emitting MLCT state, and that the protection offered by DNA versus solvent interaction restores MLCT emission.

Ru(II) complexes whose luminescence is significantly quenched in the presence of DNA have also been reported. Usually, the excited state of these species is a very good oxidant, and photoinduced reductive electron transfer involving guanine residues is responsible for the luminescence quenching [362, 369, 384]. In some cases, it has been proposed that the quenching

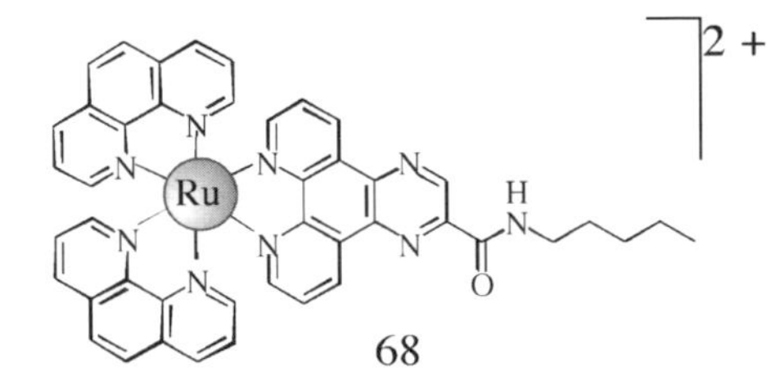

process can occur via photoinduced proton-coupled electron transfer with guanosine-5′-monophosphate.

Besides being used as luminescent probes, ruthenium complexes have been reported to form photoadducts with DNA and other species of biological relevance. The most studied photoadducts are probably the ones formed by Ru(II) complexes containing 1,4,5,8-tetraazaphenanthrene (TAP) as ligand and guanine residues on DNA strands (see for example **69**) [386]. The mechanism of photoadduct formation has been extensively investigated. The initially formed MLCT state undergoes reductive electron transfer from guanine. This process is followed by fast formation of a covalent bond between the electron donor and acceptor, which leads to an adduct between the metallic complex and the nucleobase [386]. Such photoadduct formation has also been used to induce photocrosslinks between two nucleotide strands when one of the strands was chemically derivatized by the photoreactive metal complex and the complementary strand contained a guanine base in the proximity of the tethered complex [387]. The necessary requirement for this photoreaction to occur is a MLCT excited state which is a very strong oxidant, as guaranteed by the TAP ligand.

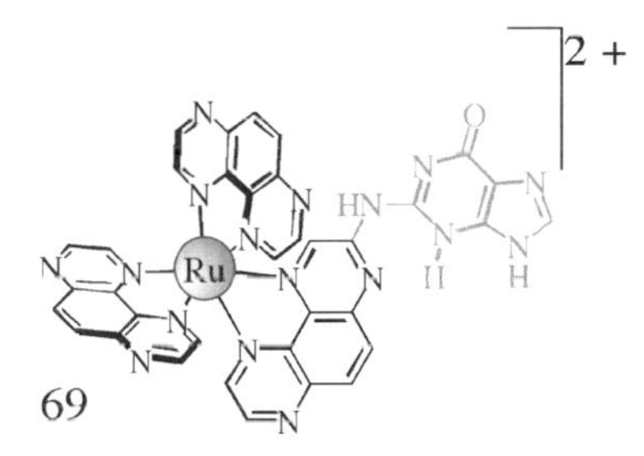

More recently, a photoadduct between similar Ru complexes and the amino acid tryptophan have also been reported [388]. The authors mention that this photoreaction appears very promising for a wide range of applications to peptides and proteins.

Ru(II) complexes have also been inserted into synthetic oligonucleotides to obtain specific information on the properties of DNA strands and/or to prepare particular (super)structures [389–393]. For example, Ru(II)-derivatized oligonucleotides have been used to investigate the distance dependence of the quenching of suitable Ru luminescence by guanine residues [393]. Oligonucleotide conjugates containing Ru(II) polypyridine units as photosensitizers have also been reported to induce photodamage on single-stranded DNA sites [394].

The potential of Ru(II)-derivatized oligonucleotides has been explored to synthesize novel, interesting, and beautiful nanometer-sized luminescent structures in which the DNA strands act as templates and the Ru complexes act as both template and photoactive units [395–397], giving rise to

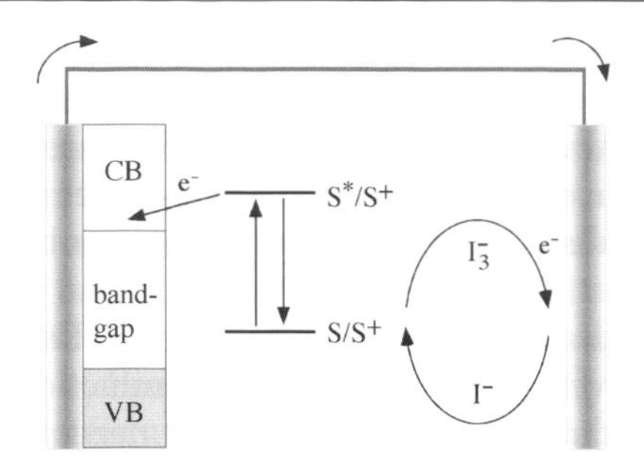

Fig. 19 Schematic operation principle of a dye-sensitized solar cell

semiconductor oxide obtained via a sol–gel procedure. Dye coverage of semiconductor nanoparticles is generally obtained from alcoholic solutions of the sensitizer, in which the sintered film is left immersed for a few hours. Sensitizers are usually designed to have functional groups such as $-COOH$, $-PO_3H_2$, or $-B(OH)_2$ for stable adsorption onto the semiconductor substrate. The dye-covered film is in intimate contact with an electrolytic solution containing a redox couple dissolved in a suitable solvent. The electron donor member of the redox couple must reduce quickly and quantitatively the oxidized sensitizer, so closing the circuit. A variety of solvents with different viscosity and of redox mediators have been the object of intense studies, the most commonly used being the couple I_3^-/I^- in acetonitrile or methoxypropionitrile solution. The counter electrode is a conductive glass covered with a few clusters of metallic platinum, which has a catalytic effect in the reduction process of the electron mediator. Further details on the cells and on their preparation can be found in the literature [403–405].

The complete photoelectrochemical cycle of the device can be outlined as follows. The adsorbed sensitizer molecules (S) are brought into their excited state (S*) by photon absorption and inject one electron into the empty conduction band of the semiconductor in a timescale of femtoseconds. Injected electrons percolate through the nanoparticle network and are collected by the conductive layer of the photoanode electrode, while the oxidized sensitizer (S^+) in its ground state is rapidly reduced by I^- ions in solution. Photoinjected electrons flow in the external circuit where useful electric work is produced and are available at the counter electrode for the reduction of the electron mediator acceptor I_3^-. The entire cycle consists in the quantum conversion of photons to electrons.

$$S + h\nu \rightarrow S^* \qquad \text{photoexcitation} \tag{25}$$

$$S^* + TiO_2 \rightarrow S^+ + (e^-, TiO_2) \qquad \text{electron injection} \tag{26}$$

$$2S^+ + 3I^- \rightarrow 2S + I_3^- \qquad \text{sensitizer regeneration} \tag{27}$$

$$I_3^- + 2e^- \rightarrow 3I^- \qquad \text{electron donor regeneration}\,. \tag{28}$$

Photoinjected electrons should escape from any recombination process in order to have a unit charge collection efficiency at the photoelectrode back contact. The two major waste processes in a dye-sensitized solar cell are due to (1) back electron transfer, at the semiconductor/electrolyte interface, between electrons in the conduction band and the oxidized dye molecules (Eq. 29), and (2) reduction of the electron relay (I_3^-, in this case) at the semiconductor nanoparticle surface (Eq. 30).

$$S^+ + (e^-, TiO_2) \rightarrow S \quad \text{back electron transfer} \quad (29)$$

$$I_3^- + 2(e^-, TiO_2) \rightarrow 3I^- \quad \text{electron capture from mediator}\,. \quad (30)$$

A detailed knowledge of all the kinetic mechanisms occurring in a photoelectrochemical cell under irradiation is an essential feature toward optimization of the process.

7.2 Ruthenium-Sensitized Photoelectrochemical Solar Cells

A major breakthrough in the field relied on the performance of dye-sensitized solar cells employing Ru(II) complexes as sensitizers [401–405, 409–411]. Several reasons are at the basis of the success of Ru(II) polypyridine complexes in playing this leading role:

1. Strong absorption throughout all the visible region, which can also extend to the near-IR. This result is obtained by means of intense MLCT bands due to a judicious choice and combination of ligands [1].
2. Strong electronic coupling between the MLCT excited state of the chromophore and the semiconductor conduction band. To fulfill this requirement, it has to be noted that the polypyridine ligand connected to the semiconductor via suitable functionalization of the ligand (usually carboxylated ligands) must be that involved in the lowest-lying MLCT state.
3. Tunability of the excited-state redox properties. This allows the preparation of compounds whose excited-state oxidation potential can ensure an efficient electron injection in the semiconductor conduction band. In this regard, it should be considered that to estimate a "reduction potential" (E_{cb}) for the semiconductor conduction band is not an easy task [404, 412–414], and in nonaqueous solvents adsorption of cations, which are present as electrolytes, also has a significant effect on E_{cb} values. For example, E_{cb} for nanostructured TiO_2 has been reported to be – 1.0 V vs SCE in 0.1M $LiClO_4$/acetonitrile and about – 2.0 V when Li^+ cations are replaced by tetrabutylammonium [413, 414].
4. Stability of the Ru(II) polypyridine complexes, in the ground state as well as in the excited and redox states. However, it is useful to note that photostability is not a strict requisite here, since the excited state is rapidly deactivated by electron injection. The same applies to chromophores hav-

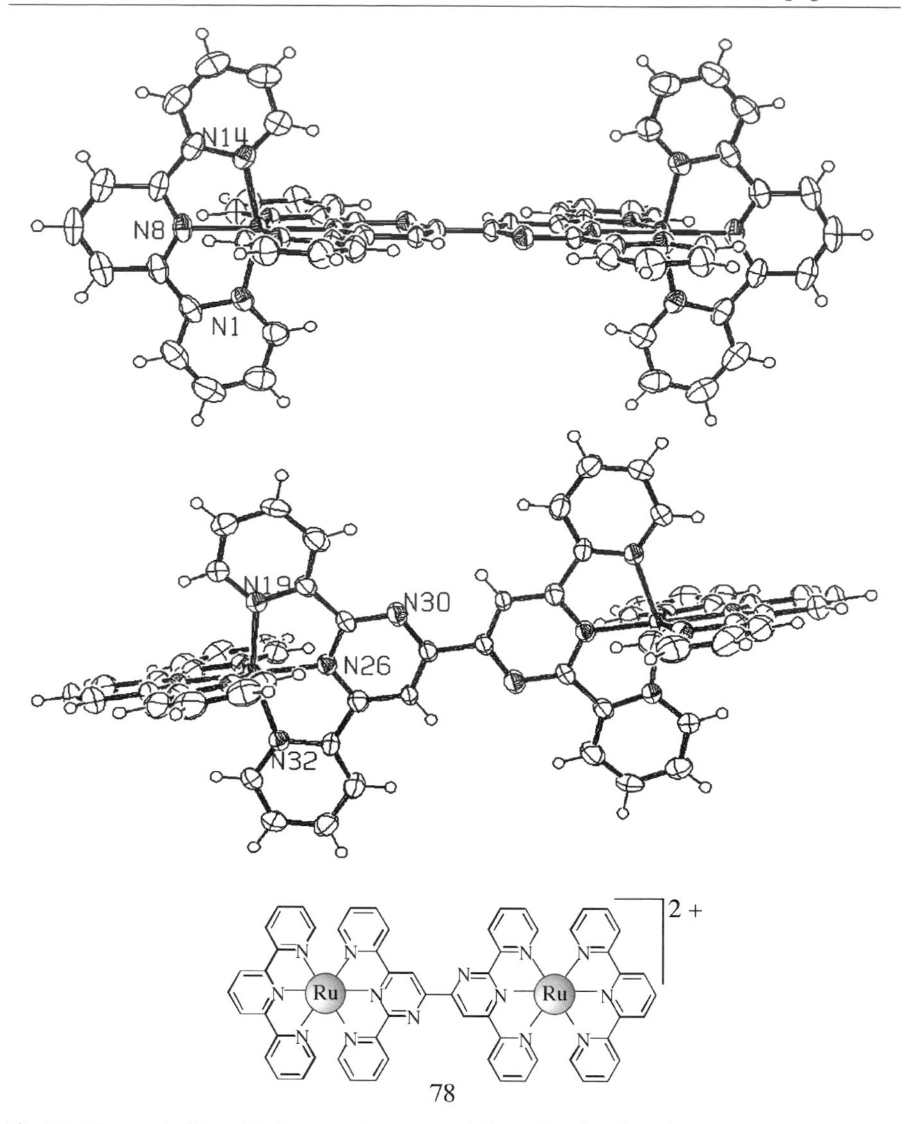

Fig. 23 Thermal ellipsoid views and structural formula of a dinuclear Ru(II) complex

Multiple emissions from Ru(II) polypyridine complexes have been reported. Besides multiple emission connected to supramolecular species featuring nonquantitative interchromophoric energy transfer (a relatively common and in some way an expected behavior), in some cases multiple emission from the same Ru(II) subunit has also been proposed at room temperature [453]. This behavior has not been fully explained yet.

Many efforts have aimed to take advantage of the photophysical properties of Ru(II) chromophores in electropolymerized thin film structures and

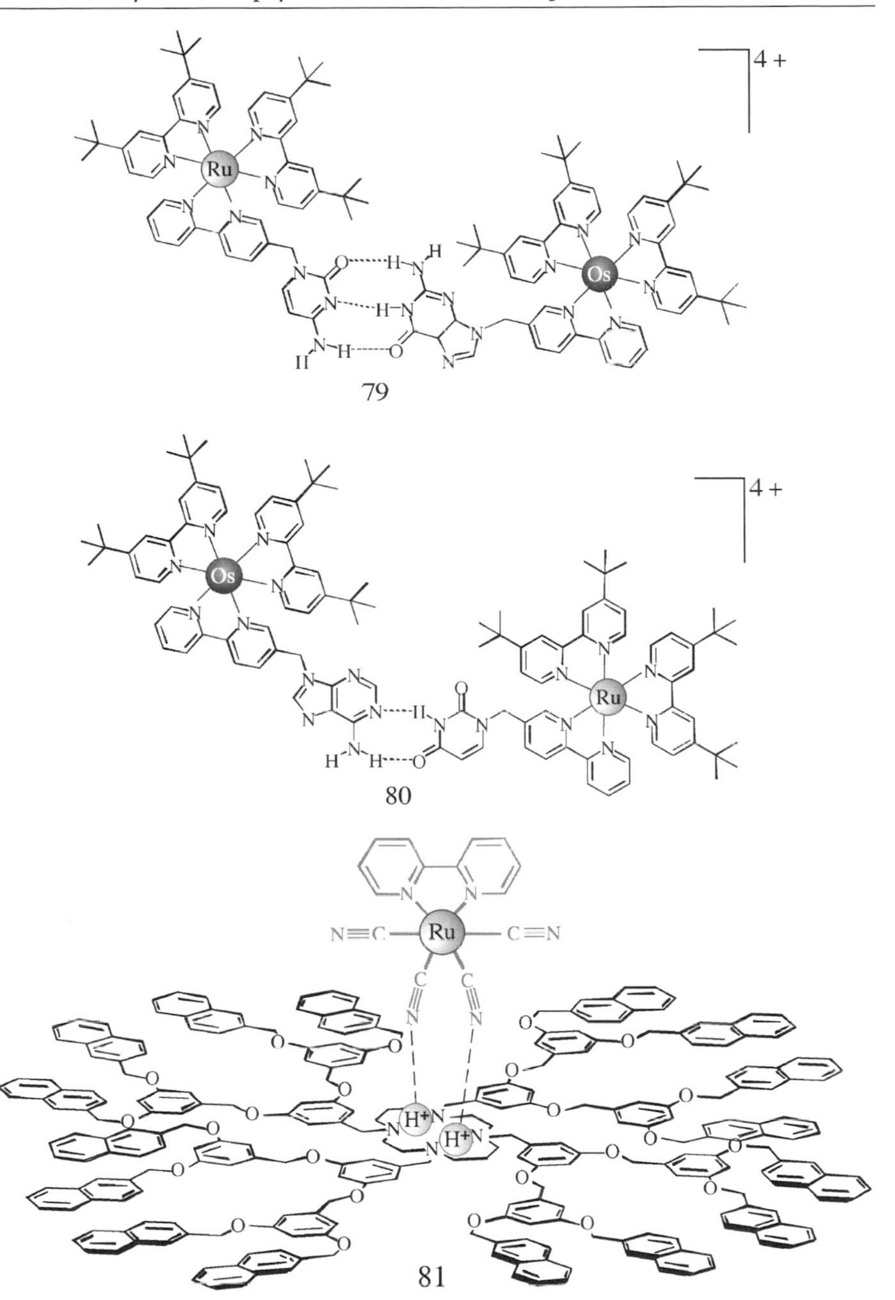

in SiO_2-based sols–gels [342]. In some cases, these solid-interfaced systems allowed interesting photocatalytic results to be obtained, such as (1) the dehydrogenation of 2-propanol to acetone and molecular hydrogen, obtained in a photoelectrochemical cell in which a dinuclear Ru complex is adsorbed

50. Sprouse S, King KA, Spellane PJ, Watts RJ (1984) J Am Chem Soc 106:6647
51. Maestri M, Sandrini D, Balzani V, Chassot L, Jolliet P, von Zelewsky A (1985) Chem Phys Lett 122:375
52. Roundhill DM (1994) Photochemistry and photophysics of metal complexes. Plenum, New York
53. Maestri M, Balzani V, Deuschel-Cornioley C, von Zelewsky A (1991) Adv Photochem 17:1
54. King KA, Spellane PJ, Watts RJ (1985) J Am Chem Soc 107:1431
55. Hillis JE, De Armond MK (1971) J Lumin 4:273
56. Shaw JR, Schmehl RH (1991) J Am Chem Soc 113:389
57. Guglielmo G, Ricevuto V, Giannetto A, Campagna S (1989) Gazz Chim Ital 119:457
58. Juris A, Campagna S, Bidd I, Lehn JM, Ziessel R (1988) Inorg Chem 27:4007
59. Ohsawa Y, Sprouse S, King KA, DeArmond MK, Hanck KW, Watts RJ (1987) J Phys Chem 91:1047
60. Maestri M, Sandrini D, Balzani V, von Zelewsky A, Jolliet P (1988) Helv Chim Acta 71:134
61. Van Houten J, Watts RJ (1976) J Am Chem Soc 98:4853
62. Van Houten J, Watts RJ (1978) Inorg Chem 17:3381
63. Elfring WH Jr, Crosby GA (1981) J Am Chem Soc 103:2683
64. Durham B, Caspar JV, Nagle JK, Meyer TJ (1982) J Am Chem Soc 104:4803
65. Caspar JV, Sullivan BP, Kober EM, Meyer TJ (1982) Chem Phys Lett 91:91
66. Calvert JM, Caspar JV, Binstead RA, Westmoreland TD, Meyer TJ (1982) J Am Chem Soc 104:6620
67. Agnew SF, Stone ML, Crosby GA (1982) Chem Phys Lett 85:57
68. Caspar JV, Meyer TJ (1983) J Am Chem Soc 105:5583
69. Barigelletti F, Juris A, Balzani V, Belser P, von Zelewsky A (1983) Inorg Chem 22:3335
70. Juris A, Barigelletti F, Balzani V, Belser P, von Zelewsky A (1985) Inorg Chem 24:1758
71. Barigelletti F, Belser P, von Zelewsky A, Juris A, Balzani V (1985) J Phys Chem 89:3680
72. Wacholtz WF, Auerbach RA, Schmehl RH (1986) Inorg Chem 25:227
73. Anderson S, Constable EC, Seddon KR, Turp JE, Baggott JE, Pilling MJ (1985) J Chem Soc Dalton Trans, p 2247
74. Barigelletti F, Juris A, Balzani V, Belser P, von Zelewsky A (1986) J Phys Chem 90:5190
75. Barigelletti F, Juris A, Balzani V, Belser P, von Zelewsky A (1987) J Phys Chem 91:1095
76. Kitamura N, Sato M, Kim HB, Obata R, Tazuke S (1988) Inorg Chem 27:651
77. Hiraga T, Kitamura N, Kim HB, Tazuke S, Mori N (1989) J Phys Chem 93:2940
78. Hammarström L, Barigelletti F, Flamigni L, Indelli MT, Armaroli N, Calogero G, Guardigli M, Sour A, Collin JP, Sauvage JP (1997) J Phys Chem A 101:9061
79. Abrahamson M, Wolpher H, Johansson O, Larsson J, Kritikos M, Eriksson L, Norrby PO, Bergquist J, Sun L, Akermark B, Hammarström L (2005) Inorg Chem 44:3215
80. Cherry WR, Henderson LJ Jr (1984) Inorg Chem 23:983
81. Van Houten J, Watts RJ (1975) J Am Chem Soc 97:3843
82. Englman R, Jortner J (1970) Mol Phys 18:145
83. Gelbart WM, Freed KF, Rice SA (1970) J Chem Phys 52:2460
84. Barbara P, Meyer TJ, Ratner MA (1996) J Phys Chem 100:13148
85. Gleria M, Minto F, Beggiato G, Bortolus P (1978) J Chem Soc Chem Commun, p 285
86. Porter GB, Sparks RH (1980) J Photochem 13:123

87. Fetterolf ML, Offen HW (1985) J Phys Chem 89:3320
88. Fetterolf ML, Offen HW (1986) J Phys Chem 90:1828
89. De Cola L, Barigelletti F, Balzani V, Belser P, von Zelewsky A, Vögtle F, Ebmeyer F, Grammenudi S (1988) J Am Chem Soc 110:7210
90. Barigelletti F, De Cola L, Balzani V, Belser P, von Zelewsky A, Vögtle F, Ebmeyer F (1989) J Am Chem Soc 111:4662
91. Treadway JA, Strouse GF, Anderson PA, Keene RF, Meyer TJ (2002) J Chem Soc Dalton Trans, p 3820
92. Medlycott EA, Hanan GS (2005) Chem Soc Rev 34:133
93. Juris A, Campagna S, Balzani V, Gremaud G, von Zelewsky A (1988) Inorg Chem 27:3652
94. Schofield ER, Collin JP, Gruber N, Sauvage JP (2003) Chem Commun, p 188
95. Bolletta F, Juris A, Maestri M, Sandrini D (1980) Inorg Chim Acta 44:L175
96. Laurence GS, Balzani V (1974) Inorg Chem 13:2976
97. Gafney HD, Adamson AW (1972) J Am Chem Soc 94:8238
98. Demas JN, Adamson AW (1973) J Am Chem Soc 95:5159
99. Bock CR, Meyer TJ, Whitten DG (1974) J Am Chem Soc 96:4710
100. Navon G, Sutin N (1974) Inorg Chem 13:2159
101. Hoffman MZ, Moggi L, Bolletta F, Hugh GL (1989) J Phys Chem Ref Data 18:219
102. Natarajan P, Endicott JF (1973) J Phys Chem 77:971
103. Natarajan P, Endicott JF (1973) J Phys Chem 77:1823
104. Sabbatini N, Scandola MA, Carassiti V (1973) J Phys Chem 77:1307
105. Sabbatini N, Scandola MA, Balzani V (1974) J Phys Chem 78:541
106. Juris A, Gandolfi MT, Manfrin MF, Balzani V (1976) J Am Chem Soc 98:1047
107. Ballardini R, Varani G, Scandola F, Balzani V (1976) J Am Chem Soc 98:7432
108. Juris A, Manfrin MF, Maestri M, Serpone N (1978) Inorg Chem 17:2258
109. Fujita I, Kobayashi H (1970) J Chem Phys 52:4904
110. Fujita I, Kobayashi H (1972) J Chem Phys 59:2902
111. Fujita I, Kobayashi H (1972) Ber Bunsenges Phys Chem 76:115
112. Sabbatini N, Balzani V (1972) J Am Chem Soc 94:7857
113. Bolletta F, Maestri M, Moggi L, Balzani V (1975) J Chem Soc Chem Commun, p 901
114. Creutz C, Sutin N (1976) J Am Chem Soc 98:6384
115. Creutz C (1978) Inorg Chem 17:1046
116. Sutin N, Creutz C (1978) Adv Chem Ser 168:1
117. Brunschwig B, Sutin N (1978) J Am Chem Soc 100:7568
118. Hoselton MA, Lin C-T, Schwarz HA, Sutin N (1978) J Am Chem Soc 100:2383
119. Balzani V, Bergamini G, Campagna S, Puntoriero F (2007) Photochemistry and Photophysics of Coordination Compounds: Overview and General Concepts (in this volume)
120. Balzani V, Bolletta F, Ciano M, Maestri M (1983) J Chem Educ 60:447
121. Balzani V, Bolletta F (1983) Comments Inorg Chem 2:211
122. Gafney HD, Adamson AW (1975) J Chem Educ 52:480
123. Bolletta F, Bonafede S (1986) Pure Appl Chem 58:1229
124. Liu DK, Brunschwig BS, Creutz C, Sutin N (1986) J Am Chem Soc 108:1749
125. Bolletta F, Ciano M, Balzani V, Serpone N (1982) Inorg Chim Acta 62:207
126. White HS, Bard AJ (1982) J Am Chem Soc 104:6891
127. Jonah CD, Matheson MS, Meisel D (1978) J Am Chem Soc 100:1449
128. Hercules LM, Lytle FE (1966) J Am Chem Soc 88:4795
129. Rubinstein I, Bard AJ (1981) J Am Chem Soc 103:512

200. Amouyal E, Homsi A, Chambron JC, Sauvage JP (1990) J Chem Soc Dalton Trans, p 1841
201. Friedman AE, Chambron JC, Sauvage JP, Turro NJ, Barton JK (1990) J Am Chem Soc 118:4960
202. Tuite E, Lincoln P, Norden B (1997) J Am Chem Soc 119:239
203. Nair RB, Cullum BM, Murphy CJ (1997) Inorg Chem 36:962
204. Olson EJC, Hu D, Hormann A, Jonkman AM, Arkin MR, Stemp EDA, Barton JK, Barbara PF (1997) J Am Chem Soc 119:11458
205. Albano G, Belser P, De Cola L, Gandolfi MT (1999) Chem Commun, p 1171
206. Flamigni L, Encinas S, MacDonnell FM, Kim KJ, Puntoriero F, Campagna S (2000) Chem Commun, p 1185
207. Torieda H, Nozali K, Yoshimura A, Ohno T (2004) J Phys Chem A 108:4819
208. Balzani V, Moggi L, Scandola F (1987) In: Balzani V (ed) Supramolecular photochemistry. Reidel, Dordrecht, p 1
209. Balzani V, Scandola F (1991) Supramolecular photochemistry. Horwood, Chichester
210. Chiorboli C, Indelli MT, Scandola F (2005) Top Curr Chem 257:63
211. Schlike B, Belser P, De Cola L, Sabbioni E, Balzani V (1999) J Am Chem Soc 121:4207
212. Welter S, Salluce N, Belser P, Groeneveld M, De Cola L (2005) Coord Chem Rev 249:1360
213. Indelli MT, Chiorboli C, Flamigni L, De Cola L, Scandola F (2007) Inorg Chem (in press)
214. Cavazzini M, Pastorelli P, Quici S, Loiseau F, Campagna S (2005) Chem Commun, p 5266
215. Barigelletti F, Flamigni L, Collin JP, Sauvage JP (1997) Chem Commun, p 333
216. Collin JP, Gavina P, Hietz V, Sauvage JP (1998) Eur J Inorg Chem 1
217. Barigelletti F, Flamigni L (2000) Chem Soc Rev 29:1
218. Barigelletti F, Flamigni L, Guardigli M, Juris A, Beley M, Chodorowski-Kimmes S, Collin JP, Sauvage JP (1996) Inorg Chem 35:136
219. Polson M, Chiorboli C, Fracasso S, Scandola F (2007) Photochem Photobiol Sci 6:438
220. Indelli MT, Scandola F, Collin JP, Sauvage JP, Sour A (1996) Inorg Chem 25:303
221. Indelli MT, Scandola F, Flamigni L, Collin JP, Sauvage JP, Sour A (1997) Inorg Chem 26:4247
222. Helms A, Heiler D, McLendon G (1991) J Am Chem Soc 113:4325
223. Onuchic JN, Beratan DN (1987) J Am Chem Soc 109:6771
224. Lainé PP, Bedioui F, Loiseau F, Chiorboli C, Campagna S (2006) J Am Chem Soc 128:7510
225. Lainé PP, Loiseau F, Campagna S, Ciofini I, Adamo C (2006) Inorg Chem 45:5538
226. Grosshenny V, Harriman A, Ziessel R (1995) Angew Chem Int Ed Engl 34:1100
227. Paddon-Row MN (2001) In: Balzani V (ed) Electron transfer in chemistry, vol 3. Wiley-VCH, Weinheim, p 179
228. Scandola F, Chiorboli C, Indelli MT, Rampi MA (2001) In: Balzani V (ed) Electron transfer in chemistry, vol. 3. Wiley-VCH, Weinheim, p 337
229. Grosshenny V, Harriman A, Ziessel R (1995) Angew Chem Int Ed Engl 34:2705
230. Ziessel R, Hissler M, El-ghayoury A, Harriman A (1998) Coord Chem Rev 177:1251
231. Belser P, Dux R, Baak M, De Cola L, Balzani V (1995) Angew Chem Int Ed Engl 34:595
232. De Cola L, Balzani V, Dux R, Baak M, Belser P (1995) Supramol Chem 5:297
233. Chiorboli C, Rodgers MAJ, Scandola F (2003) J Am Chem Soc 125:483
234. Welter S, Laforet F, Cecchetto E, Vergeer F, De Cola L (2005) ChemPhysChem 6:2417

235. Torieda H, Yoshimura A, Nozaki K, Sakai S, Ohno T (2002) J Phys Chem A 106:11034
236. Torieda H, Yoshimura A, Nozaki K, Ohno T (2004) J Phys Chem A 108:2148
237. Ruben M, Rojo J, Romero-Salguero FJ, Uppadine LH, Lehn JM (2004) Angew Chem Int Ed 43:3644
238. Credi A, Balzani V, Campagna S, Hanan GS, Arana CR, Lehn JM (1995) Chem Phys Lett 243:105
239. Bassani DM, Lehn JM, Serroni S, Puntoriero F, Campagna S (2003) Chem Eur J 9:5936
240. Stadler AM, Puntoriero F, Campagna S, Kyritsakas N, Welter R, Lehn JM (2005) Chem Eur J 11:3997
241. Juris A (2003) Annu Rep Prog Chem Sect C 99:177
242. Balzani V, Ceroni P, Maestri M, Saudan C, Vicinelli V (2003) Top Curr Chem 228:159
243. Balzani V, Vögtle F (2003) CR Chim 6:867
244. Balzani V, Juris A, Puntoriero F, Campagna S (2007) In: Gleria M, De Jaeger R (eds) Inorganic polymers, chap 19. Nova Science, New York
245. Balzani V, Campagna S, Denti G, Serroni S, Juris A, Venturi M (1998) Acc Chem Res 31:26
246. Pleovets M, Vögtle F, De Cola L, Balzani V (1999) New J Chem 23:63
247. Vögtle F, Pleovets M, Nieger M, Azzellini GC, Credi A, De Cola L, De Marchis V, Venturi M, Balzani V (1999) J Am Chem Soc 121:6290
248. McClenaghan ND, Passalacqua R, Loiseau F, Campagna S, Verheyde B, Hameurlaine A, Dehaen W (2003) J Am Chem Soc 125:5356
249. Zhou XL, Tyson DS, Castellano FN (2000) Angew Chem Int Ed 39:4301
250. Tyson DS, Luman CR, Castellano FN (2002) Inorg Chem 41:3578
251. Serroni S, Denti G, Campagna S, Juris A, Ciano M, Balzani V (1992) Angew Chem Int Ed Engl 31:1493
252. Campagna S, Denti G, Serroni S, Juris A, Venturi M, Ricevuto V, Balzani V (1995) Chem Eur J 1:211
253. Serroni S, Juris A, Venturi M, Campagna S, Resino Resino I, Denti G, Credi A, Balzani V (1997) J Mater Chem 7:1227
254. Denti G, Campagna S, Serroni S, Ciano M, Balzani V (1992) J Am Chem Soc 114:2944
255. Campagna S, Denti G, Sabatino L, Serroni S, Ciano M, Balzani V (1990) Inorg Chem 29:4750
256. Serroni S, Campagna S, Denti G, Keyes TE, Vos JG (1996) Inorg Chem 35:4513
257. Campagna S, Denti G, Serroni S, Ciano M, Balzani V (1991) Inorg Chem 30:3728
258. Denti G, Serroni S, Campagna S, Ricevuto V, Juris A, Ciano M, Balzani V (1992) Inorg Chim Acta 198–200:507
259. Denti G, Campagna S, Sabatino L, Serroni S, Ciano M, Balzani V (1990) Inorg Chim Acta 176:175
260. Campagna S, Denti G, Serroni S, Juris A, Ciano M, Balzani V (1991) Inorg Chem 31:2982
261. Leveque J, Moucheron C, Kirsch-De Mesmaeker A, Loiseau F, Serroni S, Puntoriero F, Campagna S, Nierengarten H, Van Dorsselaer A (2004) Chem Commun, p 877
262. Campagna S, Denti G, Sabatino L, Serroni S, Ciano M, Balzani V (1989) J Chem Soc Chem Commun, p 1500
263. Sommovigo M, Denti G, Serroni S, Campagna S, Mingazzini C, Mariotti C, Juris A (2001) Inorg Chem 40:3318
264. Puntoriero F, Serroni S, Galletta M, Juris A, Licciardello A, Chiorboli C, Campagna S, Scandola F (2005) ChemPhysChem 6:129

265. Tachibana Y, Moser JE, Grätzel M, Klug DR, Durrant JR (1996) J Phys Chem 100:20056
266. Benkö G, Kallioinen J, Korppi-Tommola JEI, Yartsev AP, Sundström V (2002) J Am Chem Soc 124:489
267. Kallioinen J, Benkö G, Sundström V, Korppi-Tommola JEI, Yartsev AP (2002) J Phys Chem B 106:4396
268. Tachibana Y, Nazeeruddin MDK, Grätzel M, Klug DR, Durrant JR (2002) Chem Phys 285:127
269. Asbury JB, Anderson NA, Hao EC, Ai X, Lian TQ (2003) J Phys Chem 107:7376
270. Benkö G, Kallioinen J, Myllyperkiö P, Trif F, Korppi-Tommola JEI, Yartsev AP, Sundström V (2004) J Phys Chem B 108:2862
271. Wenger B, Grätzel M, Moser JE (2005) J Am Chem Soc 127:12150
272. McClenaghan ND, Loiseau F, Puntoriero F, Serroni S, Campagna S (2001) Chem Commun, p 2634
273. Larsen J, Puntoriero F, Pascher T, McClenaghan ND, Campagna S, Sundström V, Åkesson E (2007) Chem Phys Chem (in press)
274. Leveque J, Elias B, Moucheron C, Kirsch-De Mesmaeker A (2005) Inorg Chem 44:393
275. Börje A, Köthe O, Juris A (2002) J Chem Soc Dalton Trans, p 843
276. Constable EC, Handel RW, Housecroft CE, Farràn Morales A, Ventura B, Flamigni L, Barigelletti F (2005) Chem Eur J 11:4024
277. Arm KJ, Williams JAG (2005) Chem Commun, p 230
278. Campagna S, Serroni S, Puntoriero F, Loiseau F, De Cola L, Kleverlaan CJ, Becher J, Sørensen AP, Hascoat P, Thorup N (2002) Chem Eur J 8:4461
279. Puntoriero F, Nastasi F, Cavazzini M, Quici S, Campagna S (2007) Coord Chem Rev 251:536
280. Balzani V, Credi A, Venturi M (2003) Molecular devices and machines. Wiley-VCH, Weinheim
281. Schanze KS, Walters KA (1998) In: Ramamurthy V, Schanze KS (eds) Organic and inorganic photochemistry. Dekker, New York, p 75
282. Collin JP, Guillerez S, Sauvage JP, Barigelletti F, De Cola L, Flamigni L, Balzani V (1991) Inorg Chem 30:4230
283. Danielson E, Elliott CM, Merkert JW, Meyer TJ (1987) J Am Chem Soc 109:2519
284. Cooley LF, Larson SL, Elliott CM, Kelley DF (1991) J Phys Chem 95:10694
285. Larson SL, Elliott CM, Kelley DF (1995) J Phys Chem 99:6530
286. Klumpp T, Linsenmann M, Larson SL, Limoges BR, Buerssner D, Krissinel B, Elliott CM, Steiner UE (1999) J Am Chem Soc 121:4092
287. Klumpp T, Linsenmann M, Larson SL, Limoges BR, Buerssner D, Krissinel B, Elliott CM, Steiner UE (1999) J Am Chem Soc 121:1076
288. Weber JM, Rawls MT, MacKenzie VJ, Limoges BR, Elliott CM (2007) J Am Chem Soc 129:313
289. Larson SL, Elliott CM, Kelley DF (1996) Inorg Chem 35:2070
290. Schmehl RH, Ryu CK, Elliott CM, Headford CLE, Ferrere S (1990) Adv Chem Ser 226:211
291. Cooley LF, Headford CEL, Elliott CM, Kelley DF (1988) J Am Chem Soc 112:6673
292. Ryu CK, Wang R, Schmehl RH, Ferrere S, Ludwikow M, Merkert JW, Headford CEL, Elliott CM (1992) J Am Chem Soc 114:430
293. Yonemoto EH, Saupe GB, Schmehl RH, Hubig SM, Riley RL, Iverson BL, Mallouk TE (1994) J Am Chem Soc 116:4786
294. Meckenburg SL, Peek BM, Schoonover JR, McCafferty DJ, Wall CJ, Erickson BW, Meyer TJ (1993) J Am Chem Soc 115:5479

295. Meckenburg SL, McCafferty DJ, Schoonover JR, Peek BM, Erickson BW, Meyer TJ (1994) Inorg Chem 33:2974
296. Hu YZ, Tsukiji S, Shinkai S, Oishi S, Hamachi I (2000) J Am Chem Soc 122:241
297. Huynh MHV, Dattelbaum DM, Meyer TJ (2005) Coord Chem Rev 249:457
298. McCafferty DG, Bishop BM, Wall CG, Hughes SG, Meckenburg SL, Meyer TJ, Erickson BW (1995) Tetrahedron 51:1093
299. McCafferty DG, Friesen DA, Danielson E, Wall CG, Saederholm MJ, Erickson BW, Meyer TJ (1993) Proc Natl Acad Sci USA 93:8200
300. Striplin DR, Reece SY, McCafferty DG, Wall CG, Friesen DA, Erickson BW, Meyer TJ (2004) J Am Chem Soc 126:5282
301. Slate CA, Striplin DR, Moss JA, Chen PY, Erickson BW, Meyer TJ (1998) J Am Chem Soc 120:4885
302. Serron SA, Aldridge SA III, Fleming CN, Danell RM, Baik MH, Sykora M, Dattelbaum DM, Meyer TJ (2004) J Am Chem Soc 126:14506
303. Serron SA, Aldridge SA III, Danell RM, Meyer TJ (2000) Tetrahedron 41:4039
304. Strouse GF, Worl LA, Younathan JN, Meyer TJ (1989) J Am Chem Soc 111:9101
305. Jones WE Jr, Baxter SM, Strouse GF, Meyer TJ (1993) J Am Chem Soc 115:7363
306. Worl LA, Jones WE Jr, Strouse GF, Younathan JN, Danielson E, Maxwell KA, Sykora M, Meyer TJ (1999) Inorg Chem 28:2705
307. Dupray LM, Devenny M, Striplin DR, Meyer TJ (1997) J Am Chem Soc 119:10243
308. Fleming CN, Dupray LM, Papanikolas JM, Meyer TJ (2002) J Phys Chem 106:2328
309. Worl LA, Strouse GF, Younathan JN, Baxter SM, Meyer TJ (1990) J Am Chem Soc 112:7571
310. Baxter SM, Jones WE Jr, Danielson E, Worl LA, Strouse GF, Younathan JN, Meyer TJ (1991) Coord Chem Rev 111:47
311. Sykora M, Maxwell KA, DeSimone JM, Meyer TJ (2000) Proc Natl Acad Sci USA 97:7687
312. Konduri R, Ye H, MacDonnell FM, Serroni S, Campagna S, Rajeshwar K (2002) Angew Chem Int Ed 41:3185
313. Chiorboli C, Fracasso S, Scandola F, Campagna S, Serroni S, Konduri R, MacDonnell FM (2003) Chem Commun, p 1658
314. Chiorboli C, Fracasso S, Ravaglia M, Scandola F, Campagna S, Wouters K, Konduri R, MacDonnell FM (2005) Inorg Chem 44:1658
315. Konduri R, de Tacconi NR, Rajeshwar K, MacDonnell FM (2004) J Am Chem Soc 126:11621
316. de Tacconi NR, Lezna RO, Konduri R, Ongeri F, Rajeshwar K, MacDonnell FM (2005) Chem Eur J 11:4327
317. Wouters KL, de Tacconi NR, Konduri R, Lezna RO, MacDonnell FM (2006) Photosynth Res 87:1
318. Molnar SM, Nallas G, Bridgewater JS, Brewer KJ (1994) J Am Chem Soc 116:5206
319. Holder AA, Swavey S, Brewer KJ (2004) Inorg Chem 43:303
320. Elvington M, Brewer KJ (2006) Inorg Chem 45:5242
321. Barber J (2003) Q Rev Biophys 36:71
322. Ferreira KN, Iverson TM, Maghlaoui K, Barber J, Iwata S (2004) Science 303:1831
323. Iwata S, Barber J (2004) Curr Opin Struct Biol 14:447
324. McEvoy JP, Brudvig G (2006) Chem Rev 106:4455
325. Hammarström L (2003) Curr Opin Chem Biol 7:666
326. Berg KE, Tran A, Raymond MK, Abrahamsson M, Wolny J, Redon S, Andersson M, Sun L, Styring S, Hammarström L, Toflund H, Akermark B (2001) Eur J Inorg Chem 1019

327. Abrahamsson M, Baudin HB, Tran A, Philouze C, Berg KE, Raymond MK, Sun L, Akermark B, Styring S, Hammarström L (2002) Inorg Chem 41:1534
328. Magnusson A, Frapart Y, Abrahamsson M, Horner O, Akermark B, Sun L, Girerd JJ, Hammarström L, Styring S (1999) J Am Chem Soc 121:89
329. Burdinski D, Wieghardt K, Steenken S (1999) J Am Chem Soc 121:10781
330. Sjodin M, Styring S, Sun L, Akermark B, Hammarström L (2000) J Am Chem Soc 122:3932
331. Sun L, Raymond MK, Magnusson A, LeGourriérec D, Tamm M, Abrahamsson M, Huang-Kenez P, Martensson J, Stenhagen G, Hammarström L, Akermark B, Styring S (2000) J Inorg Biochem 78:15
332. Burdinski D, Eberhard B, Wieghardt K (2000) Inorg Chem 39:105
333. Huang P, Magnusson A, Lomoth R, Abrahamsson M, Tamm M, Sun L, van Rotterdam B, Park J, Hammarström L, Akermark B, Styring S (2002) J Inorg Biochem 91:150
334. Borgström M, Shaikh N, Johansson O, Anderlund MF, Styring S, Akermark B, Magnusson A, Hammarström L (2005) J Am Chem Soc 127:17504
335. Balzani V, Moggi L, Manfrin MF, Bolletta F, Gleria M (1975) Science 189:852
336. Lehn JM, Sauvage JP (1977) Nouv J Chim 1:449
337. Kalyanasundaram K, Grätzel M (1979) Angew Chem Int Ed Engl 18:701
338. Creutz C, Sutin N (1975) Proc Natl Acad Sci USA 72:2858
339. Kalyanasundaram K, Kiwi J, Grätzel M (1978) Helv Chim Acta 61:2720
340. Meyer TJ (1989) Acc Chem Res 22:163
341. Campagna S, Serroni S, Puntoriero F, Di Pietro C, Ricevuto V (2001) In: Balzani V (ed) Electron transfer in chemistry, vol. 5. Wiley-VCH, Weinheim, p 168
342. Alstrum-Acevedo JH, Brennaman MK, Meyer TJ (2005) Inorg Chem 44:6802
343. Armaroli N, Balzani V (2007) Angew Chem Int Ed 46:52
344. Wasielewski MR (1992) Chem Rev 92:435
345. Gust D, Moore TA, Moore AL (1993) Acc Chem Res 26:198
346. Ozawa H, Haga MA, Sakai K (2006) J Am Chem Soc 128:4926
347. Rau S, Schäfer B, Gleick D, Anders E, Rudolph M, Friedrich M, Görls H, Henry W, Vos JG (2006) Angew Chem Int Ed 45:6215
348. Brewer KJ, Elvington M (2006) US Patent 20060120954A1
349. Gholankhass B, Mametsuka H, Koite K, Tanabe T, Furue M, Ishitani O (2005) Inorg Chem 44:2326
350. Hawecker J, Lehn JM, Ziessel R (1983) J Chem Soc Chem Commun, p 536
351. Walther D, Ruben M, Rau S (1999) Coord Chem Rev 182:67
352. Inagaki A, Edure S, Yatsuda S, Akita M (2005) Chem Commun, p 5468
353. Osawa M, Hoshino M, Wakatsuki Y (2001) Angew Chem Int Ed 40:3472
354. Rau S, Walther D, Vos JG (2007) Dalton Trans, p 915
355. Ballardini R, Balzani V, Credi A, Gandolfi MT, Venturi M (2001) Acc Chem Res 34:445
356. Balzani V (2003) Photochem Photobiol Sci 2:459
357. Venturi M, Balzani V, Ballardini R, Credi A, Gandolfi MT (2004) Int J Photoenerg 6:1
358. Balzani V, Credi A, Venturi M (2007) Nano Today 2:18
359. Balzani V, Clemente-Leon M, Credi A, Ferrer B, Venturi M, Fllod A, Stoddart JF (2006) Proc Natl Acad Sci USA 103:1178
360. Mobian P, Kern J-P, Sauvage J-P (2004) Angew Chem Int Ed 43:2392
361. Collin J-P, Jouvenot D, Koizumi M, Sauvage J-P (2005) Eur J Inorg Chem 1850
362. Kirsch-De Mesmaeker A, Lecomte JP, Kelly JM (1996) Top Curr Chem 177:25

363. Jenkins Y, Friedman AE, Turro NJ, Barton JK (1992) Biochemistry 31:10809
364. Friedman AE, Chambron JC, Sauvage JP, Turro NJ, Barton JK (1984) Nucleic Acids Res 13:6016
365. Tossi AB, Kelly JM (1989) Photochem Photobiol 49:545
366. Ortmans I, Moucheron C, Kirsch-De Mesmaeker A (1998) Coord Chem Rev 168:233
367. Kelly JM, Feeney M, Jacquet L, Kirsch-De Mesmeaker A, Lecomte JP (1997) Pure Appl Chem 69:767
368. Piérard F, Del Guerzo A, Kirsch-De Mesmaeker A, Demeunynck M, Lhomme J (2001) Phys Chem Chem Phys 3:2911
369. Vos JG, Kelly JM (2006) Dalton Trans, p 4839
370. Hiort C, Lincoln P, Nordén B (1993) J Am Chem Soc 115:3448
371. Lincoln P, Broo A, Nordén B (1996) J Am Chem Soc 118:2644
372. Ujj L, Coates CG, Kelly JM, Kruger P, McGarvey JJ, Atkinson GH (2002) J Phys Chem B 106:4854
373. Coates CG, Callaghan PL, McGarvey JJ, Kelly JM, Kruger P, Higgins ME (2000) J Raman Spectrosc 31:283
374. Turro C, Bossman SH, Jenkins Y, Barton JK, Turro NJ (1995) J Am Chem Soc 117:9026
375. Sabatini E, Nikol HD, Gray HB, Anson FC (1996) J Am Chem Soc 118:1158
376. Nair RB, Cullum BM, Murphy CJ (1997) Inorg Chem 36:962
377. Guo XQ, Castellano FN, Li L, Lakowitz JR (1998) Biophys Chem 71:51
378. Brennaman MK, Alstrum-Acevedo JH, Fleming CN, Jang P, Meyer TJ, Papanikolas JM (2002) J Am Chem Soc 124:15094
379. Pourtois G, Beljonne D, Moucheron C, Schumm S, Kirsch-De Mesmaeker A, Lazzaroni R, Brédas JL (2004) J Am Chem Soc 126:683
380. Onfelt B, Lincoln P, Nordéen B, Baskin JS, Zewail AH (2000) Proc Natl Acad Sci USA 97:5708
381. Coates CG, Olofsson J, Coletti M, McGarvey JJ, Onfelt B, Lincoln P, Nordén B, Tuite E, Matousek P, Parker AW (2001) J Phys Chem B 105:12653
382. Coates CG, Callaghan P, McGarvey JJ, Kelly JM, Jacquet L, Kirsch-De Mesmaeker A (2002) J Mol Struct 598:15
383. O'Donoghue K, Kelly JM, Kruger PE (2004) Dalton Trans, p 13
384. Moucheron C, Kirsch-De Mesmaeker A, Kelly JM (1997) J Photochem Photobiol B 40:91
385. Ortmans I, Elias B, Kelly JM, Moucheron C, Kirsch-De Mesmaeker A (2004) Dalton Trans, p 668
386. Jacquet L, Kelly JM, Kirsch-De Mesmaeker A (1995) J Chem Soc Chem Commun, p 913
387. Lentzen O, Defrancq E, Constant JF, Schumm S, Garcia-Fresnadillo D, Moucheron C, Dumy P, Kirsch-De Mesmaeker A (2004) J Biol Inorg Chem 9:100
388. Gicquel E, Boisdenghien A, Defranc E, Moucheron C, Kirsch-De Mesmaeker A (2006) Chem Commun, p 2764
389. Jenkins Y, Barton JK (1992) J Am Chem Soc 114:8736
390. Hurley DJ, Tor Y (1998) J Am Chem Soc 120:2194
391. Telser J, Cruickshank A, Schanze KS, Netzel TL (1989) J Am Chem Soc 111:7221
392. Grimm GN, Boutorine AS, Lincoln P, Nordén B, Hélène C (2002) ChemBioChem 3:324
393. Lentzen O, Constant JF, Defrancq E, Prevost M, Schumm S, Moucheron C, Dumy P, Kirsch-De Mesmaeker A (2003) ChemBioChem 4:195

394. Crean CW, Kavanagh YT, O'Keefe CM, Lawler MP, Stevenson C, Davies JH, Boyle PH, Kelly JM (2002) Photochem Photobiol Sci 1:1024
395. Vargas-Baca I, Mitra D, Zulyniak HJ, Banerjee J, Sleiman HF (2001) Angew Chem Int Ed 40:4629
396. Mitra D, Di Cesare N, Sleiman HF (2004) Angew Chem Int Ed 43:5804
397. Chen B, Sleiman HF (2004) Macromolecules 37:5866
398. Pauly M, Kayser I, Schmitz M, Dicato M, Del Guerzo A, Kolber I, Moucheron C, Kirsch-De Mesmaeker A (2002) Chem Commun, p 1086
399. Ortmans I, Content S, Boutonnet N, Kirsch-De Mesmaeker A, Bannwarth W, Constant JF, Defrancq E, Lhomme J (1999) Chem Eur J 5:2712
400. Garcia-Fresnadillo D, Boutonnet N, Schumm S, Moucheron C, Kirsch-De Mesmaeker A, Defrancq E, Constant JF, Lhomme J (2002) Biophys J 82:978
401. O'Regan B, Grätzel M (1991) Nature 335:737
402. Hagfeldt A, Grätzel M (1992) Chem Rev 95:49
403. Grätzel M (2005) Inorg Chem 44:6841
404. Meyer G (2005) Inorg Chem 44:6852
405. Bignozzi CA, Argazzi R, Caramori S (2007) Inorganic and bioinorganic chemistry. In: Bertini I (ed) Encyclopaedia of life supporting systems (EOLSS), developed under the auspices of UNESCO. Eolss, Oxford (in press) http://www.eolss.net
406. Gerischer H, Tributsch H (1968) Ber Bunsenges Phys Chem 72:437
407. DeSilvestro J, Grätzel M, Kavan L, Moser JE, Augustynski J (1985) J Am Chem Soc 107:2988
408. Memming R (1984) Prog Surf Sci 17:7
409. Nazeeruddin MK, Pechy P, Renouard T, Zakeeruddin SM, Humphry-Baker R, Comte P, Liska P, Cevey L, Costa E, Shklover V, Spiccia L, Deacon GB, Bignozzi CA, Grätzel M (2001) J Am Chem Soc 123:1613
410. Nazeeruddin MK, Wang Q, Cevey L, Aranyos V, Liska P, Figgemeier E, Klein C, Hirata N, Koops S, Haque SA, Durrant JR, Hagfeldt A, Lever ABP, Grätzel M (2006) Inorg Chem 45:787
411. Grätzel M (2001) Pure Appl Chem 73:459
412. Cao F, Oskam G, Searson PC, Stipkala J, Farzhad F, Heimer TA, Meyer GJ (1995) J Phys Chem 99:11974
413. Redmond G, Grätzel M, Fitzmaurice D (1993) J Phys Chem 97:6951
414. Boschlo G, Fitzmaurice D (1999) J Phys Chem B 103:7860
415. Nazeeruddin MK, Kay A, Rodicio R, Humphry-Baker R, Muller E, Liska P, Vlachopoulos M, Grätzel M (1993) J Am Chem Soc 115:6382
416. Argazzi R, Bignozzi CA, Hasselmann GM, Meyer GJ (1998) Inorg Chem 37:4533
417. Rensmo H, Sodergen S, Patthey L, Westermak L, Vayssieres L, Kohle O, Bruhwiler PA, Hagfeldt A, Siegbahn H (1997) Chem Phys Lett 274:51
418. Nazeeruddin MK, Pechy P, Grätzel M (1997) Chem Commun, p 1705
419. Zakeeruddin SM, Nazeeruddin MK, Pechy P, Rotzinger F, Humphry-Baker R, Kalyanasundaram K, Grätzel M, Shklover V, Haibach T (1997) Inorg Chem 36:5937
420. Argazzi R, Bignozzi CA, Heimer TA, Meyer GJ (1997) Inorg Chem 36:2
421. Indelli MT, Bignozzi CA, Harriman A, Schoonover JR, Scandola F (1994) J Am Chem Soc 116:3768
422. Argazzi R, Bignozzi CA, Heimer TA, Castellano FN, Meyer GJ (1997) J Phys Chem B 101:2591
423. Hirata N, Lagref JJ, Palomares EJ, Durrant JR, Nazeeruddin MK, Grätzel M, Di Censo D (2004) Chem Eur J 10:595

424. de Silva AP, Tecilla P (eds) (2005) Thematic issue on fluorescent sensors. J Mater Chem 15:2617–2976
425. Schmittel M, Lin HW (2007) Angew Chem Int Ed 46:893
426. Demas JN, DeGraff BA (1991) Anal Chem 63:829A
427. Demas JN, DeGraff BA, Coleman P (1999) Anal Chem 71:793A
428. Demas JN, DeGraff BA (2001) Coord Chem Rev 211:317
429. Borisov SM, Vasylevska AS, Krause C, Wolfbeis O (2006) Adv Funct Mater 16:1536
430. Welter S, Brunner K, Hofstraat JW, De Cola L (2003) Nature 421:54
431. Ford WE, Rodgers MAJ (1992) J Phys Chem 96:2917
432. Wilson GJ, Launikonis A, Sasse WHF, Mau AWH (1997) J Phys Chem A 101:4860
433. Simon JA, Curry SL, Schmehl RH, Schartz TR, Piotrowiak P, Jin X, Thummel RP (1997) J Am Chem Soc 119:11012
434. Harriman A, Hissler M, Khatyr A, Ziessel R (1999) Chem Eur J 5:3366
435. Tyson DS, Lauman CR, Zhou X, Castellano FN (2001) Inorg Chem 40:4063
436. McClenaghan ND, Barigelletti F, Maubert B, Campagna S (2002) Chem Commun, p 602
437. Passalacqua R, Loiseau F, Campagna S, Fang YQ, Hanan GS (2003) Angew Chem Int Ed 42:1608
438. Leroy-Lhez S, Belin C, D'Aleo A, Williams R, De Cola L, Fages F (2003) Supramol Chem 15:627
439. Maubert B, McClenhagan ND, Indelli MT, Campagna S (2003) J Phys Chem A 107:447
440. Wang XY, Del Guerzo A, Schmehl RH (2004) J Photochem Photobiol C 5:55
441. McClenaghan ND, Leydet Y, Maubert B, Indelli MT, Campagna S (2005) Coord Chem Rev 249:1336
442. Wang J, Fang YQ, Bourget-Merle L, Polson MIJ, Hanan GS, Juris A, Loiseau F, Campagna S (2006) Chem Eur J 12:8539
443. Strouse GF, Schoonover JR, Duesing R, Boyde S, Jones WE, Meyer TJ (1995) Inorg Chem 34:473
444. Fang YQ, Taylor NJ, Hanan GS, Loiseau F, Passalacqua R, Campagna S, Nierengerten H, Van Dorsselaer A (2002) J Am Chem Soc 124:7912
445. Polson MIJ, Loiseau F, Campagna S, Hanan GS (2006) Chem Commun, p 1301
446. Fang YQ, Taylor NJ, Laverdière F, Hanan GS, Loiseau F, Nastasi F, Campagna S, Nierengerten H, Leize-Wagner E, Van Dorsselaer A (2007) Inorg Chem 46:2854
447. Encinas S, Simpson NRM, Andrews P, Ward MD, White CW, Armaroli N, Barigelletti F, Houlton A (2000) New J Chem 24:987
448. Ward MD, White CM, Barigelletti F, Armaroli N, Calogero G, Flamigni L (1998) Coord Chem Rev 171:481
449. Rau S, Schäfer B, Schebesta S, Grübing A, Poppitz W, Walther D, Duati M, Browne WR, Vos JG (2003) Eur J Inorg Chem, p 1503
450. Loiseau F, Marzanni G, Quici S, Indelli MT, Campagna S (2003) Chem Commun, p 286
451. Bergamini G, Saudan C, Ceroni P, Maestri M, Balzani V, Gorka M, Lee SK, van Heyst J, Vögtle F (2004) J Am Chem Soc 126:16466
452. Falz JA, Williams RM, Dilva MJJP, De Cola L, Pikramenou Z (2006) J Am Chem Soc 128:4520
453. Glazer EC, Madge D, Tor Y (2005) J Am Chem Soc 127:4190
454. Treadway JA, Moss JA, Meyer TJ (1999) Inorg Chem 38:4386
455. Gallaghaer LA, Serron SA, Wen X, Hornstein BJ, Dattelbaum DM, Schoonover JR, Meyer TJ (2005) Inorg Chem 44:2089

456. Sykora M, Petruska MA, Alstrum-Acevedo J, Bezel I, Meyer TJ, Klimov VI (2006) J Am Chem Soc 128:9984
457. Kim YI, Salim S, Huq MJ, Mallouk TE (1991) J Am Chem Soc 113:9561
458. Kim YI, Atherton SJ, Brigham ES, Mallouk TE (1993) J Phys Chem 97:11802
459. Saupe GB, Mallouk TE, Kim W, Schmehl RH (1997) J Phys Chem 101:2508
460. Hara, Waraksa CC, Lean JT, Lewis BA, Mallouk TE (2000) J Phys Chem A 104:5275
461. Morris ND, Suzuki M, Mallouk TE (2004) J Phys Chem A 108:9115
462. Haga MA, Ali MDM, Koseki S, Fujimoto K, Yoshimura A, Nozaki K, Ohno T, Nakajima K, Stufkens D (1996) Inorg Chem 35:3335
463. Di Pietro C, Serroni S, Campagna S, Gandolfi MT, Ballardini R, Fanni S, Browne W, Vos JG (2002) Inorg Chem 41:2871
464. Fanni S, Di Pietro C, Serroni S, Campagna S, Vos JG (2000) Inorg Chem Commun 3:42
465. Jukes TF, Adamo V, Hartl F, Belser P, De Cola L (2005) Coord Chem Rev 249:1327
466. Belser P, De Cola L, Hartl F, Adamo V, Bozic B, Chriqui Y, Iyer VM, Jukes TF, Kühni J, Querol M, Roma S, Salluce N (2006) Adv Funct Mater 16:195
467. Saes M, Bressler C, Abela R, Grolimund D, Johnson SL, Heimann PA, Chergui M (2003) Phys Rev Lett 90:47403
468. Gawelda W, Johnson M, de Groot FMF, Abela R, Bressler C, Chergui M (2006) J Am Chem Soc 128:5001
469. Benfatto M, Della Longa S, Hatada H, Hayakawa K, Gawelda W, Bressler C, Chergui M (2006) J Phys Chem B 110:14035

Top Curr Chem (2007) 280: 215–255
DOI 10.1007/128_2007_137

Published online: 27 June 2007

Photochemistry and Photophysics of Coordination Compounds: Rhodium

Maria Teresa Indelli · Claudio Chiorboli · Franco Scandola (✉)

Dipartimento di Chimica dell'Università, ISOF-CNR sezione di Ferrara, 44100 Ferrara, Italy
snf@unife.it

Abstract Rhodium(III) polypyridine complexes and their cyclometalated analogues display photophysical properties of considerable interest, both from a fundamental viewpoint and in terms of the possible applications. In mononuclear polypyridine complexes, the photophysics and photochemistry are determined by the interplay between LC and MC excited states, with relative energies depending critically on the metal coordination environment. In cyclometalated complexes, the covalent character of the C – Rh bonds makes the lowest excited state classification less clear cut, with strong mixing of LC, MLCT, and LLCT character being usually observed. In redox reactions, Rh(III) polypyridine units can behave as good electron acceptors and strong photo-oxidants. These properties are exploited in polynuclear complexes and supramolecular systems containing these units. In particular, Ru(II)-Rh(III) dyads have been actively investigated for the study of photoinduced electron transfer, with specific interest in driving force, distance, and bridging ligand effects. Among systems of higher nuclearity undergoing photoinduced electron transfer, of particular interest are polynuclear complexes where rhodium dihalo polypyridine units, thanks to their Rh(III)/Rh(I) redox behavior, can act as two-electron storage components. A large amount of work has been devoted to the use of

Rh(III) polypyridine complexes as intercalators for DNA. In this role, they have proven to be very versatile, being used for direct strand photocleavage marking the site of intercalation, to induce long-distance photochemical damage or dimer repair, or to act as electron acceptors in long-range electron transfer processes.

Keywords DNA intercalators · Electron transfer · Photophysics · Polynuclear complexes · Rhodium

Abbreviations

bpy	2,2′-bipyridine
bzq	Benzo(h)-quinoline
chrysi	5,6-chrysenequinone diimine
dpb	2,3-bis(2-pyridyl)benzoquinoxaline
DPB	4,4′-diphenylbipyridine
dpp	2,3-bis(2-pyridyl)pyrazine
dppz	dipyridophenazine
dpq	2,3-bis(2-pyridyl)quinoxaline
HAT	1,4,5,8,9,12-hexaazatriphenylene
Me_2bpy	4,4′-dimethyl-2,2′-bipyridine
Me_2phen	4,7-dimethyl-1,10-phenanthroline
Me_2trien	diamino-4,7-diazadecane
ox	Oxalato
phen	1,10-phenanthroline
phi	9,10-phenanthrenequinonediimine
PPh_3	triphenylphosphine
ppy	2-phenylpyridine,
TAP	1,4,5,8-tetraazaphenanthrene
thpy	2-(2-thienyl)-pyridine
tpy	2,2′ : 6′,2″-terpyridine

1 Introduction

Although not as popular as other transition metals, e.g., ruthenium, rhodium has received considerable attention in the field of inorganic photochemistry. Few specific reviews on rhodium photochemistry are available, however. The literature preceding 1970 was reviewed in the classical book of Carassiti and Balzani [1]. The photochemistry of polypyridine metal complexes, including those of rhodium, has been reviewed by Kalyanasundaram in 1992 [2]. Several rhodium-containing species are considered in the extensive review written in 1996 by Balzani and coworkers on luminescent and redox active polynuclear complexes [3]. A number of photochemical investigations are included in the 1997 review article of Hannon on rhodium complexes [4]. Rhodium complexes are included in more recent reviews dealing with photoinduced processes in covalently linked systems containing metal complexes [5, 6].

In general terms, three main classes of rhodium complexes have attracted the attention of photochemists: (i) amino complexes and substituted derivatives; (ii) multiply bridged dirhodium complexes; (iii) rhodium polypyridine and related complexes.

Rhodium(III) halo/amino complexes have been actively investigated in the late 1970s and in the 1980s. They have low-lying metal centered (MC) excited stats of d – d type, and can be considered paradigmatic representatives of the ligand-field photochemistry of d^6 metal complexes. The subject has been summarized and clearly discussed by Ford and coworkers in their 1983 review articles [7, 8]. In more recent times, however, despite a number of interesting investigations [9–18], the activity in the field seems to have slowed down considerably.

Multiply bridged dirhodium complexes constitute a large family of compounds, the structure and properties of which depend strongly upon the oxidation state of the metals. The Rh(I)-Rh(I) species of formula $Rh_2(bridge)_4{}^{2+}$, where the two d^8 metal centers are bridged by four bidentate ligands (e.g., diisocyanoalkanes) in square planar coordination, have $d\sigma^* \rightarrow p\sigma$ excited states with a greatly shortened metal–metal bond [19] that emit efficiently in fluid solution [20]. In the Rh(II)-Rh(II) species of formula $Rh_2(bridge)_4X_2{}^{n+}$, the two d^7 metal centers are bridged by four bidentate ligands (e.g., diisocyanoalkanes, acetate) and complete their pseudo-octahedral coordination with a metal–metal bond and two-axial monodentate ligands (e.g., X = Cl, Br $n = 2$). These dirhodium complexes have long-lived excited states of $d\pi^* \rightarrow d\sigma^*$ type, which do not emit in fluid solution but can undergo a variety of bimolecular energy and electron transfer reactions [21]. Dirhodium tetracarboxylato units of this type have also been used as building blocks for a variety of supramolecular systems of photophysical interest [22–24]. Particularly interesting triply bridged dirhodium complexes of type $X_2Rh(bridge)_3RhX_2$, $LRh(bridge)_3RhX_2$, and $LRh(bridge)_3RhL$ (bridge = bis(difluorophosphino)methylamine, X = Br, L = PPh_3) have been developed recently by Nocera [25]. These Rh(II)-Rh(II), Rh(0)-Rh(II) and Rh(0)-Rh(0) species, all possessing excited states of $d\pi^* \rightarrow d\sigma^*$ type, can be interconverted photochemically by means of two-electron redox processes. Such two-electron photoprocesses provide the basis for a recently developed light-driven hydrogen production system [26], with a Rh(0)-Rh(II) mixed valence species playing the role of key photocatalysts [27]. The interest in multiply bridged dirhodium systems is now largely driven by their potential and implications for photocatalytic purposes.

As for other transition metals, polyimine ligands (in particular, polypyridines and their cyclometalated analogues) have played a major role in the design of rhodium complexes of photophysical interest. This is due to an ensemble of factors, including chemical robustness, synthetic flexibility, electronic structure, excited-state and redox tunability. Thus, rhodium polypyridine and related complexes have been extensively studied from a photophysical viewpoint, both as simple molecular species or as components of supramolecular systems featuring energy/electron transfer processes. Also, rhodium polypyridine complexes have played a major role in the active research field of DNA metal complex interactions. In recognition of the relevance of these systems, this review will be essentially focused on the photochemistry and photophysics of complexes of rhodium with polypyridine-type ligands and of supramolecular systems that that contain such units as molecular components.

2 Mononuclear Species

2.1 Polypyridine Complexes

The fundamental features of the photophysics of Rh(III) polypyridine complexes have been extensively discussed in the book of Kalyanasundaram [2] and only a few general aspects are recalled here. The tris(1,10-phenanthroline)rhodium(III) ion, $Rh(phen)_3^{3+}$ (**1**), can be used to exemplify the typical photophysical behavior of this class of complexes. $Rh(phen)_3^{3+}$ exhibits in 77 K matrices an intense (Φ, ca. 1), long-lived (τ, ca. 50 ms), structured emission (λ = 465, 485, 524, 571 nm) assigned as ligand-centered (LC) phosphorescence, i.e., emission from a π–π^* triplet state essentially localized on the phenanthroline ligands [28–31]. As is shown by high-resolution spectroscopy, the LC excitation is not delocalized, but rather confined to a single ligand [32, 33]. In room-temperature fluid solutions, $Rh(phen)_3^{3+}$ is practically non-emitting (see below). The LC triplet state can be nevertheless easily monitored by transient absorption spectroscopy (λ_{max} = 490 nm, ε_{max} = 4000 M^{-1} cm^{-1}, τ = 250 ns) [34]. The temperature-dependent behavior is explained on the basis of decay of the LC triplet via a thermally activated process involving an upper metal-centered (MC) state [34–36]. Indeed, the

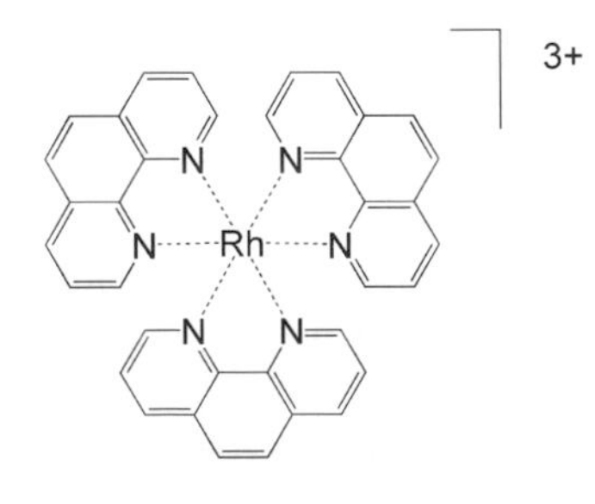

(1)

properties of the very weak emission measured from room-temperature solutions of $Rh(phen)_3{}^{3+}$ (bandshape, lifetimes) are consistent with a small amount of MC excited state being in thermal equilibrium with the lowest LC state [34]. In related 2,2′-bipyridine complexes, changes in LC-MC energy gap and emission spectral profile can be induced by methyl substitution in the 3,3′ positions, as a consequence of changes in the degree of planarity of the ligand [37].

The interplay of LC and MC triplets in the photophysics of this type of complexes has been investigated in detail by studying the series of mixed-ligand complexes *cis*-$Rh(phen)_2XY^{n+}$ ($X = Y = CN$, $n = 1$; $X = Y = NH_3$, $n = 3$; $X = NH_3$, $Y = Cl$, $n = 2$), where the energy of the MC states is controlled by the ligand field strength of the X, Y ancillary ligands [38]. While at room temperature all the complexes are very weakly emissive, at 77 K, the di-cyano and di-amino complexes give the typical, structured LC emission, whereas the amino-chloro complex exhibits a broad emission of MC type (Fig. 1a). This can be readily explained on the basis of the energy diagram of Fig. 1b, where the relative energies of the LC triplet (appreciably constant for the three complexes) and of the lowest MC state (dependent on the ligand field strength of the ancillary ligands) are schematically depicted. For the amino-chloro complex the MC state is the lowest excited state of the system. For the di-amino case the situation is similar to that of the $Rh(phen)_3{}^{3+}$ complex, with LC as the lowest excited state but with MC sufficiently close in energy to provide an efficient thermally activated decay path for LC (actually, in an appropriate temperature regime the two states are in equilibrium). In the di-cyano complex, the MC state is sufficiently high in energy that the LC state has a substantial lifetime (1.2 μs) even in room-temperature solution [38]. The actual energy gap between the LC and MC states (2000 cm^{-1} for the di-cyano

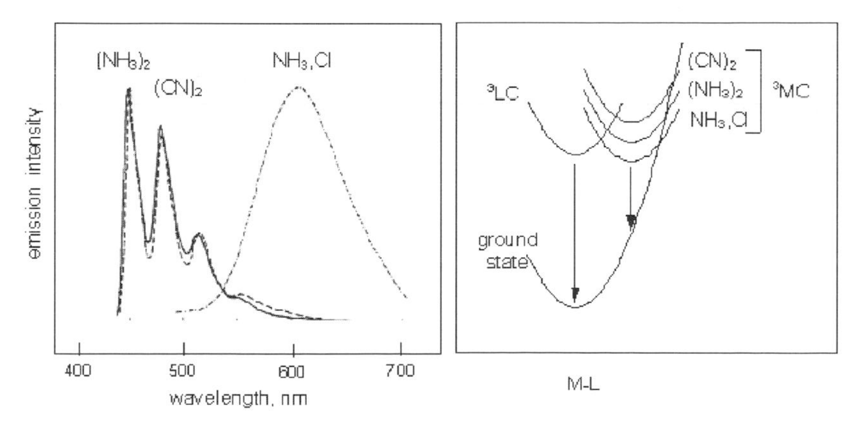

Fig. 1 **a** 77 K emission spectra of *cis*-$Rh(phen)_2XY^{n+}$ complexes with different X, Y ancillary ligands. **b** Rationalization of the emission properties in terms of relative energies of ligand-centered (LC) and metal-centered (MC) excited states (adapted from [38])

cases ligand-to-metal charge transfer (LMCT) photochemistry is observed. A clear example is provided by the $Rh(bpy)_2(ox)^+$ complex (**6**) [52].

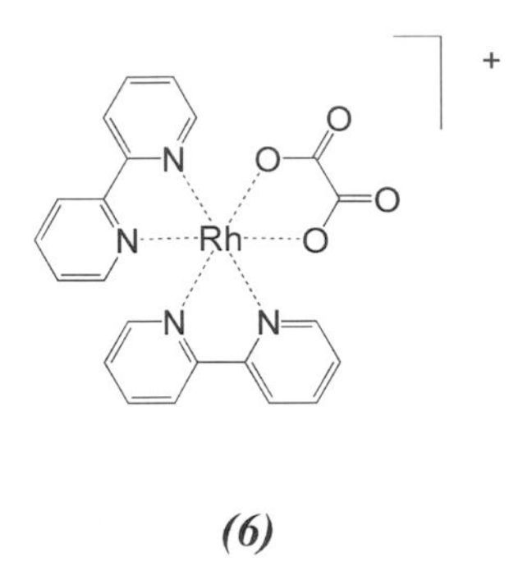

(6)

The spectrum of the colorless complex **6** is characterized by an intense band at ca. 300 nm assigned to oxalato-to-rhodium LMCT transitions. Upon UV irradiation, the following photoreaction is observed:

$$Rh^{III}(ox)(bpy)_2{}^+ + h\nu \rightarrow Rh^{I}(bpy)_2{}^+ + 2CO_2\,. \tag{1}$$

The $Rh(bpy)_2{}^+$ product is formed rapidly (risetime in pulsed experiments, ca. 10 ns), probably via a sequence of processes comprising photochemical intramolecular electron transfer from the oxalate ligand to Rh(III) followed by the decomposition of the oxidized ligand into CO_2 and $CO_2{}^-$ radical (Eq. 2) and thermal reduction of the Rh(II) center to Rh(I) by the reactive $CO_2{}^-$ radical (Eq. 3) [52].

$$Rh^{III}(ox)(bpy)_2{}^+ + h\nu \rightarrow Rh^{II}(CO_2{}^-)(bpy)_2{}^+ + CO_2 \tag{2}$$

$$Rh^{II}(CO_2{}^-)(bpy)_2{}^+ \rightarrow Rh^{I}(bpy)_2{}^+ + CO_2\,. \tag{3}$$

The violet $Rh(bpy)_2{}^+$ product, with intense MLCT visible absorption, is a tetrahedrally distorted d^8 square planar complex [53]. This Rh(I) species, which can also be obtained by chemical [54], electrochemical [55], or radiation chemical [56, 57] reduction of Rh(III) complexes, is of great interest from the catalytic viewpoint. It undergoes facile oxidative addition by molecular hydrogen [58], to give the corresponding Rh(III) dihydride (Eq. 4). The reaction is fully reversible upon

$$Rh(bpy)_2{}^+ + H_2 \leftrightarrows cis\text{-}Rh^{III}(bpy)_2(H)_2{}^+ \tag{4}$$

removal of molecular hydrogen from the system. Interestingly, the release of molecular hydrogen from the dihydride complex can be obtained photochemically (Eq. 5). This photoreaction provides a

$$cis\text{-}Rh^{III}(bpy)_2(H)_2 \xrightarrow{h\nu} Rh(bpy)_2{}^+ + H_2 \tag{5}$$

convenient means to perturb the equilibrium of Eq. 4 and to study the kinetics of hydrogen uptake in the thermal relaxation of the perturbed system. A detailed picture of the transition state of this interesting reaction has been obtained by such type of experiments [53].

2.2 Cyclometalated Complexes

Ligands such as 2-phenylpyridine, 2-(2-thienyl)-pyridine, or benzo(h)-quinoline, in their ortho-deprotonated forms (ppy, **7**; thpy, **8**; bzq, **9**), can bind in a bidentate N^C fashion to a variety of transition metals, including Rh(III). These complexes are generally indicated as cyclometalated (or orthometalated) complexes. The spectroscopy and photophysics of cyclometalated complexes [59] are usually very different form those of the corresponding polypyridine complexes, the main reason being the much stronger σ donor character of C^- relative to N. The consequences are (i) a high degree of covalency in the carbon–metal bond, (ii) a strongly enhanced ease of oxidation of the metal, (iii) high-energy MC excited states, (iv) relatively low-energy MLCT excited states.

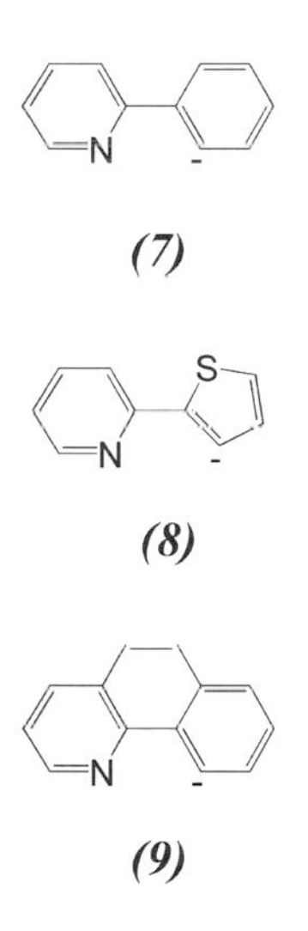

(7)

(8)

(9)

The photophysics of Rh(III) cyclometalated complexes, though not as developed as that of analogous Ir(III) species (see Chap. 9), has been actively investigated in the last two decades. While some tris- [60] and mono-cyclometalated [61] compounds have been synthesized and studied, for synthetic reasons bis-cyclometalated complexes of Rh(III) are by far more common in the literature. All the bis-cyclometalated complexes have a C,C *cis* geometry [62]. The $Rh(ppy)_2(bpy)^+$ (**10**) complex can be used here to exemplify the main photophysical features of this class of compounds.

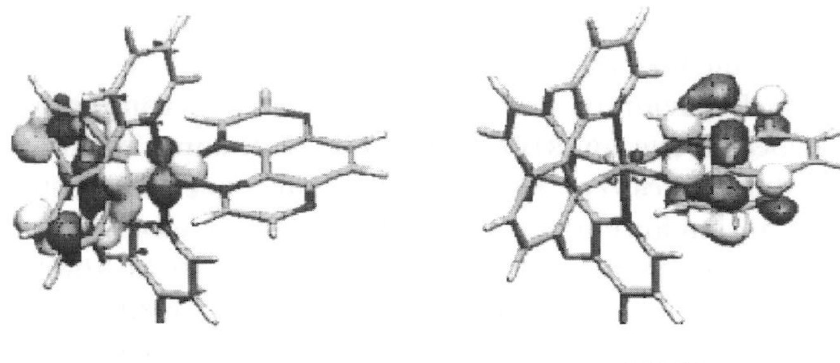

Fig. 3 Frontier orbitals of $Rh(ppy)_2TAP^+$. From [67]

3 Polynuclear and Supramolecular Species

3.1 Homobinuclear Complexes

A few homobinuclear ligand-bridged Rh(III) polypyridine complexes have been studied [3, 75, 76]. Their photochemical interest is rather limited, however, as they behave generally like their mononuclear analogues, with minor differences in spectral shifts and lifetimes. An interesting type of systems, which bring together the complexities of ligand-bridged species and multiply bridged metal–metal bonded rhodium dimers, has recently been reported by Campagna et al. [77]. In (**14**) two quadruply bridged Rh(II)-Rh(II) dimers are the "molecular components" of a higher-order two-component system held together by the complex bis-naphthyridine-type ligand. As already observed for some related simple rhodium dimers (e.g., $Rh_2(CH_3COO)_4(PPh_3)_2$) [21], the "binuclear" compound has a non-emissive but long-lived excited state in room-temperature solution. In the case of **14**, the long-lived state has been assigned as an MLCT (metal–metal π^* to naphthyridine π^*) excited state [77].

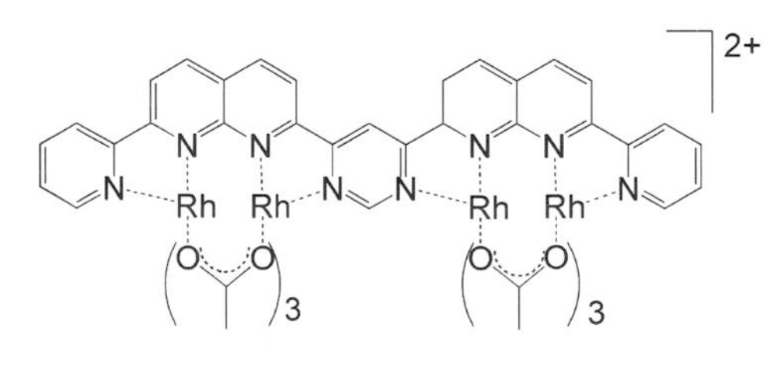

(14)

3.2 Dyads

Heteronuclear bimetallic species containing rhodium polypyridine complexes are more interesting, as the Rh(III) unit can be involved in intercomponent processes, particularly of the electron transfer type. The thermodynamic requirements for the participation of Rh(III) polypyridine complexes in electron transfer processes are summarized in Fig. 4, where Rh(III), *Rh(III), and Rh(II) represent the ground state, the triplet LC excited state (see Sect. 2.1), and the one-electron reduced form, respectively, and the values of excited-state energy [31] and reduction potential [78] refer to $Rh(phen)_3{}^{3+}$ (**1**). From these figures, it is apparent that Rh(III) polypyridine complexes can behave as extremely powerful photochemical oxidants and relatively good electron transfer quenchers. On the other hand, because of the high excited-state energy, these complexes are also good potential energy donors.

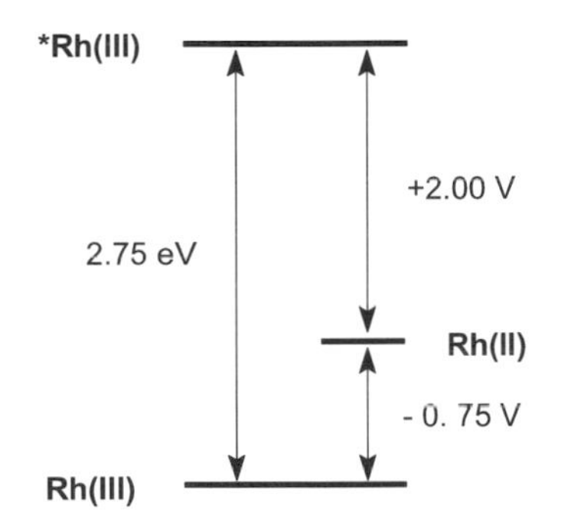

Fig. 4 Typical redox energy level diagram for Rh(III) polypyridine complexes. Values (reduction potential vs. SCE) appropriate for $Rh(phen)_3{}^{3+}$ (**1**)

As a matter of fact, Rh(III) polypyridine complexes have been extensively used in bimolecular electron transfer processes, either as photoexcited molecule [79, 80] or as quencher [81–83], with motivations of both fundamental (testing electron transfer-free energy relationships) [80] and applied nature (photoinduced hydrogen evolution from water) [81, 82]. Here, on the other hand, we focus our attention on photoinduced processes where the reactants are pre-assembled in some kind of supramolecular system. The most common photoinduced processes taking place in simple two-component systems (often called "*dyads*") involving a Rh(III) polypyridine unit are shown in Eqs. 6–8:

$$^*\mathrm{Rh(III)} - \mathrm{Q} \rightarrow \mathrm{Rh(III)} - {}^*\mathrm{Q} \quad (6)$$

$$^*\mathrm{Rh(III)} - \mathrm{Q} \rightarrow \mathrm{Rh(II)} - \mathrm{Q}^+ \quad (7)$$

$$^*\mathrm{P} - \mathrm{Rh(III)} \rightarrow \mathrm{P}^+ - \mathrm{Rh(II)}\,. \quad (8)$$

detailed kinetic picture of Fig. 6 [84]. The widely different rates of the three ET processes (a, b, d) can be rationalized in terms of predominant driving force effects [84], as shown schematically in Fig. 7.

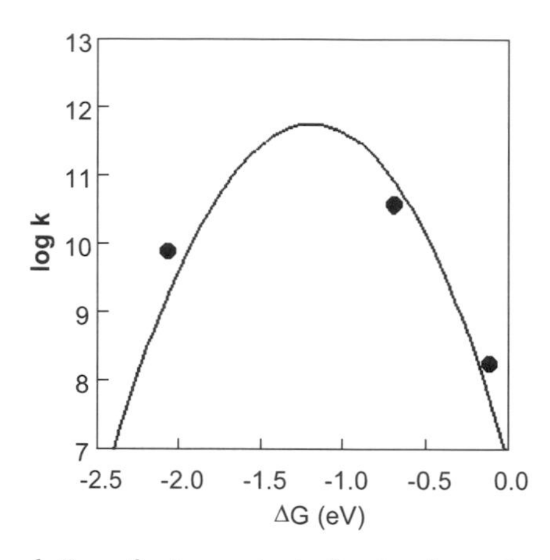

Fig. 7 Free-energy correlation of rate constants for the three electron transfer processes of dyad **15**

Since most Ru(II)-Rh(III) polypyridine dyads have very similar energetics, the qualitative features illustrated above for dyad **15** can be safely generalized to this whole class of compounds. For example, in dyad **16** [85] the rate constants of processes a, b, and d are slower by a factor of ca. 3 but have the same *relative* magnitudes as for dyad **15**. The slower rates are likely related to the longer aliphatic bridge, although for this and related [86] dyads, the flexibility of the bridges limits the validity of such comparisons.

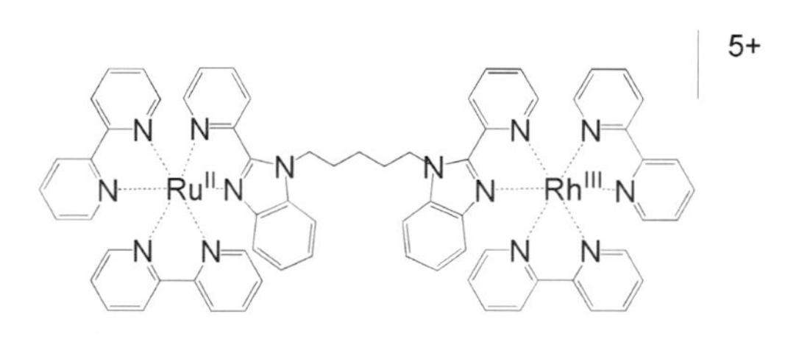

(16)

Within this general type of behavior of Ru(II)-Rh(III) dyads, a number of experimental studies have been specifically aimed at investigating the role of the bridge in determining electron transfer rates. As has been the case for other types of bimetallic dyads [87–90], particular attention has been devoted to Ru(II)-Rh(III) dyads with modular bridges involving p-phenylene spacer units [6, 91, 92]. The dyads in Chart 1 provide a homogeneous series with appreciably constant energetics but differing in the number (1–3) of *p*-phenylene spacers in the bridge and, in one of the two dyads with three spacers, for the presence of alkyl substituents on the central spacer. Upon excitation of the Ru(II)-based chromophore, the rate constants of photoinduced electron transfer (process a in the above general scheme) have been measured by time-resolved emission and transient absorption techniques in the nanosecond and picosecond time domains [6, 92]. The values for **Ru-ph-Rh**, **Ru-ph$_2$-Rh**, and **Ru-ph$_3$-Rh** (Chart 1), when plotted as a function of the metal–metal distance r (Fig. 8), display an exponential decrease (Eq. 9):

$$k = k(0)\exp(-\beta r)\,. \tag{9}$$

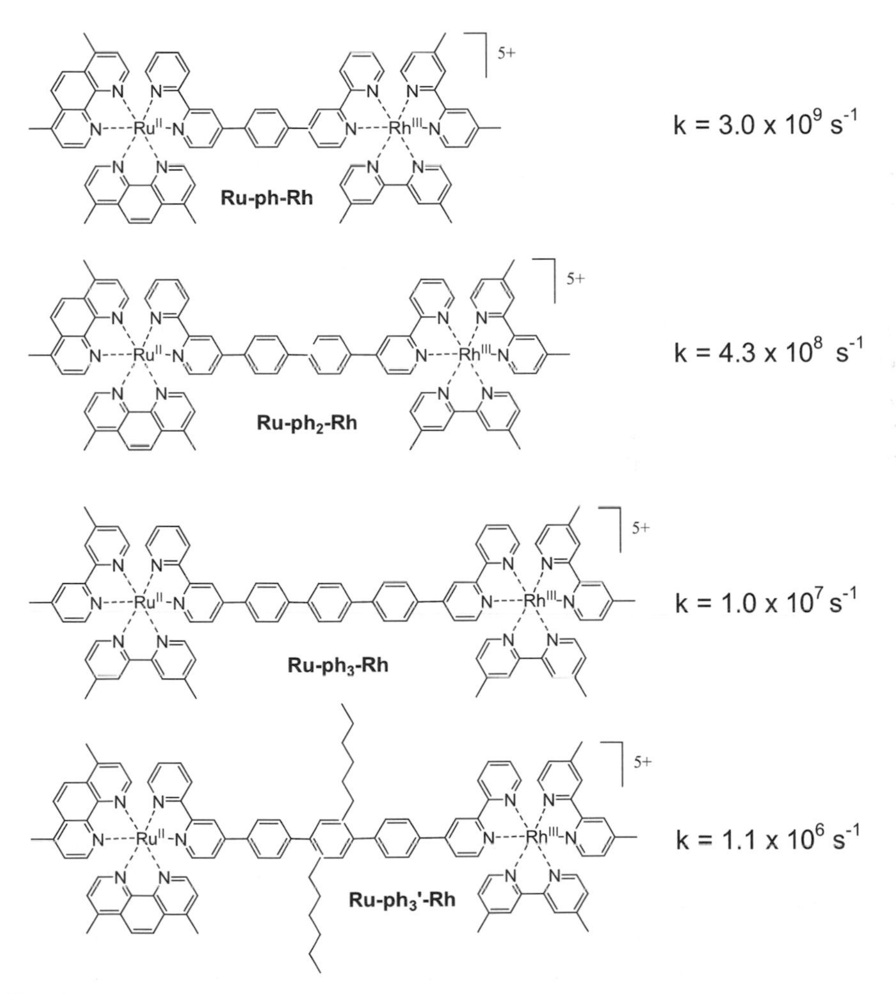

Chart 1

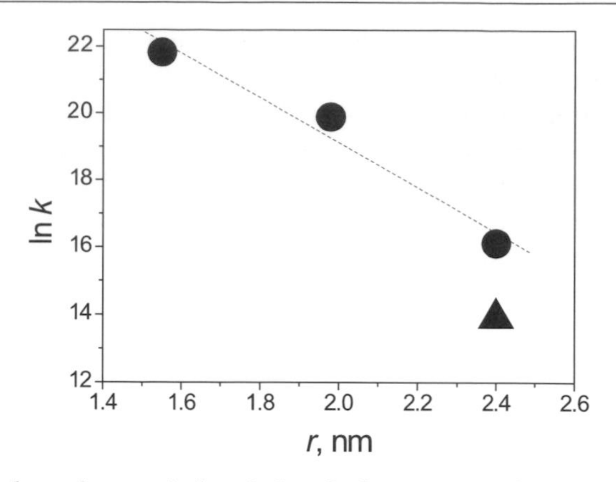

Fig. 8 Distance dependence of photoinduced electron transfer rates in the dyads of Chart 1: **Ru-ph-Rh, Ru-ph₂-Rh, Ru-ph₃-Rh** (*dots*), **Ru-ph′₃-Rh** (*triangle*)

This is the behavior predicted for electron transfer in the superexchange regime [5,93,95] if the distance dependence of the reorganizational energy term can be neglected. The β value obtained from the slope of the line in Fig. 8, 0.65 Å^{-1}, should be regarded as an upper limiting value for the attenuation factor of the intercomponent electronic coupling (Eq. 9). This β value is in the range found for other oligophenylene-containing systems (organic dyads [96, 97], metal–molecules–metal junctions [98]). This underlines the good ability of this type of bridges to mediate donor–acceptor electronic coupling (for comparison, β is typically 0.8–1.2 Å^{-1} for rigid aliphatic bridges). In this regard, it is instructive to compare the electron transfer rate constant observed for **Ru-ph-Rh** ($k = 3.0 \times 10^9$ s^{-1}) with that mentioned above for dyad **15** containing an aliphatic bis-methylene bridge ($k = 1.7 \times 10^8$ s^{-1}). Despite the longer metal–metal distance (15.5 Å for **Ru-ph-Rh** relative to 13.5 Å for **15**), the reaction is faster across the phenylene spacer by more than one order of magnitude.

An interesting result [6, 92] is the fact that dyad **Ru-ph′₃-Rh**, which is identical to **Ru-ph₃-Rh** except for the presence of two solubilizing hexyl groups on the central phenylene ring, is one order of magnitude slower than its unsubstituted analog (Fig. 8). This is related to the notion that in a superexchange mechanism the rate is sensitive to the electronic coupling between adjacent modules of the spacer [5, 93, 95], and that in polyphenylene bridges this coupling is a sensitive function of the twist angle between adjacent spacers [99]. While the planes of unsubstituted adjacent phenylene units form angles of. 20°–40° [100, 101], ring substitution leads to a substantial increase in the twist angle (to ca. 70°) [100] and, as a consequence, to a slowing down of the electron transfer process.

A number of Ru(II)-Rh(III) dyads have been reported where little or any photoinduced electron transfer quenching of the Ru(II)-based MLCT emission takes place [75, 76, 102]. In the case of dyad **17**, the plausible reason is that, owing to the comparable reduction potentials of the diphenylpyrazine bridging ligand and Ru(III) center, the driving force for intramolecular electron transfer is too small [102]. In the cases of dyads **18** and **19**, the presence of cyclometalated ancillary ligands makes the formally Rh(III) center very difficult to reduce and relatively easy to oxidize, thus yielding MLCT states at comparable energies on the two units [75, 76].

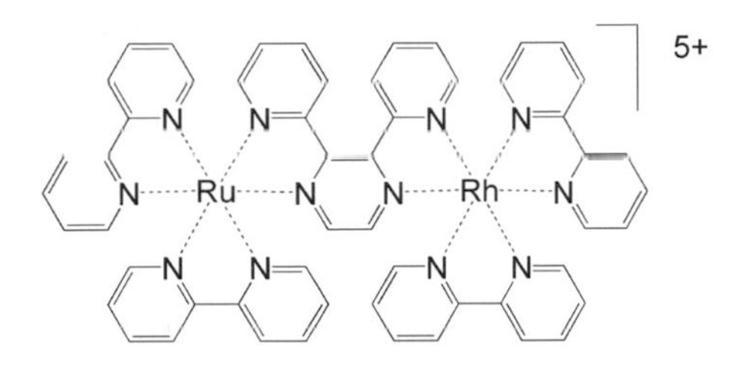

(17)

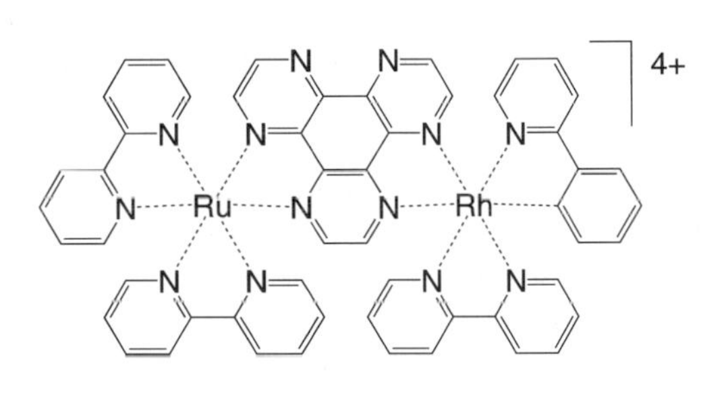

(18)

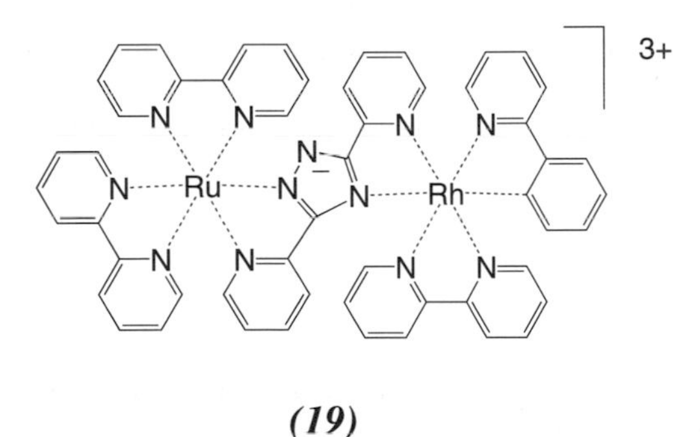

(19)

Low driving force arguments could also apply to the dyad **20**, where relatively slow quenching of the Ru(II) MLCT emission (estimated k, ca. $3.5 \times 10^7\ s^{-1}$) was observed and attributed to intramolecular electron transfer [103]. Here, however, a relevant aspect is also the presence a Rh(III)

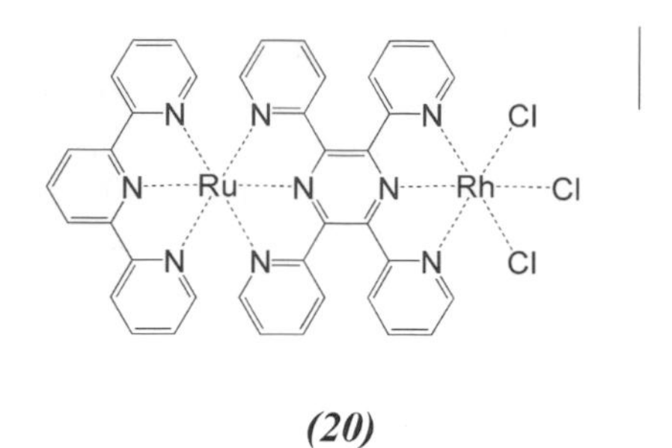

(20)

moiety with of coordinated chloride ligands. In fact, contrary to what happens for common Rh(III) polypyridine units (e.g., $Rh(bpy)_3{}^{3+}$, $Rh(phen)_3{}^{3+}$) where one-electron reduction is a quasi-reversible process, mixed-ligand units containing halide ions (e.g., $Rh(bpy)_2Cl_2{}^{+}$, $Rh(phen)_2Cl_2{}^{+}$) undergo strongly irreversible two-electron reductions accompanied by prompt halide ligand loss [78, 104]. While the use of these units as electron acceptors can be of interest towards photoinduced electron collection and multi-electron catalysis (see Sect. 3.4), from a kinetic viewpoint the large reorganizational energies involved are likely to lead to slow electron transfer rates.

3.2.2
Photoinduced Electron Transfer in Porphyrin-Rh(III) Conjugates

Though structurally quite different, the porphyrin-Rh(III) conjugates thoroughly studied by Harriman et al. [105] behave with regard to photoinduced electron transfer rather similarly to the above-discussed Ru(II)-Rh(III) dyads. The systems contain a zinc porphyrin unit connected directly with one (**21**) or via a phenylene spacer with two (**22**) rhodium terpyridine units.

The electron transfer processes thermodynamically allowed in these systems are indicated in Eqs. 10–14, where both **21** and **22** are schematized as

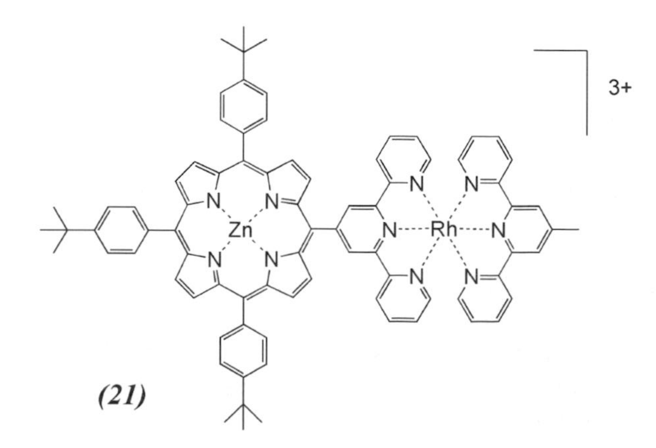

(21)

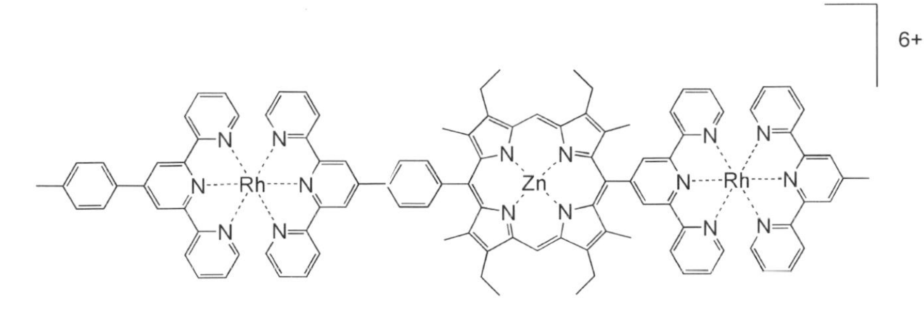

(22)

dyads (indeed, **22** is a triad only in a formal sense), ZnP, Rh, and ph represent the zinc porphyrin and rhodium terpyridine molecular components and the phenylene spacer, respectively. The singlet excited state is considered for the zinc porphyrin chromophore and the triplet state for the rhodium terpyridine unit.

$$^{*}\text{ZnP-ph-Rh(III)} \rightarrow \text{ZnP}^{+}\text{-ph-Rh(II)} \qquad \Delta G^{\circ} = -0.71\ \text{eV} \qquad (10)$$

$$\text{ZnP-ph-Rh(III)}^{*} \rightarrow \text{ZnP}^{+}\text{-ph-Rh(II)} \qquad \Delta G^{\circ} = -0.81\ \text{eV} \qquad (11)$$

$$\text{ZnP}^{+}\text{-ph-Rh(II)} \rightarrow \text{ZnP-ph-Rh(III)} \qquad \Delta G^{\circ} = -1.47\ \text{eV} \qquad (12)$$

$$^{*}\text{ZnP-Rh(III)} \rightarrow \text{ZnP}^{+}\text{-Rh(II)} \qquad \Delta G^{\circ} = -0.58\ \text{eV} \qquad (13)$$

$$\text{ZnP-Rh(III)}^{*} \rightarrow \text{ZnP}^{+}\text{-Rh(II)} \qquad \Delta G^{\circ} = -0.80\ \text{eV} \qquad (14)$$

In **22**, where a phenylene spacer is interposed between the two molecular components, both photoinduced electron transfer following excitation of the zinc porphyrin (Eq. 10) and charge recombination (Eq. 12) have been time resolved, with values in acetonitrile of $3.2 \times 10^{11}\ \text{s}^{-1}$ and $8.3 \times 10^{9}\ \text{s}^{-1}$, respectively. The wide difference in rates is attributed to the different kinetic regimes of the two processes, photoinduced electron transfer (Eq. 10) being almost activationless, and charge recombination (Eq. 12) lying deep into the Marcus inverted region [105]. For the directly linked system **21** (as well as for its free-base analogue), the disappearance of the porphyrin excited state, presumably by photoinduced electron transfer (Eq. 13), is extremely fast. In fact, the excited state lifetime of **21**, ca. 0.7 ps in acetonitrile, is comparable to the longitudinal relaxation time of the solvent, implying that the electron transfer process in this system is controlled by solvent reorientation [105].

3.3 Triads and Other Complex Systems

In dyads, including those discussed in the previous sections, photoinduced electron transfer is always followed by fast charge recombination. This greatly limits the use of dyads for practical purposes such as, e.g., conversion of light

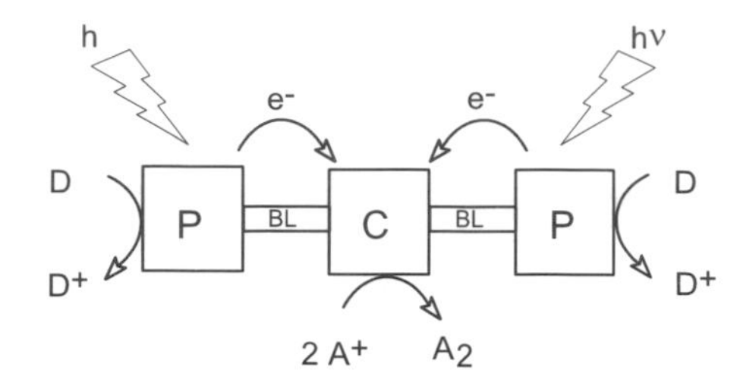

Fig. 12 Block diagram representation of a photochemical molecular device for photoinduced electron collection and two-electron redox catalysis

nents capable of storing electrons and using them in multielectron redox processes. A possible PEC scheme is shown in Fig. 12.

In this scheme, P are electron transfer photosensitizers, C an electron store component, and BL rigid bridging ligands. D is a sacrificial electron donor, and A_2 is a two-electron reduced product (e.g., H_2 starting from $2H^+$). The key molecular component C must have the ability to store two electrons following photoinduced electron transfer from P, and to deliver them to the substrate A^+ in a low-activation two-electron process that leads to the desired product A_2.

Very few homogeneous systems for photoinduced electron collection (PEC) have been reported by now [115–122]. Trimetallic complexes incorporating polyazine bridging ligands have been designed and studied by Brewer's group for potential applications as PEC devices [123–126]. Most of these complexes have general formula $[(bpy)_2Ru(BL)]_2MCl_2{}^{n+}$ with BL = 2,3-bis(2-pyridyl)pyrazine (dpp), 2,3-bis(2-pyridyl)quinoxaline (dpq), or 2,3-bis(2-pyridyl)benzoquinoxaline (dpb) and M = Ir(III) or Rh(III) [123, 125]. The first functioning PEC system, $[(bpy)_2Ru(dpb)]_2IrCl_2{}^{5+}$, was reported in 1994, employing π systems of polyazine bridging ligands to collect electrons [123]. Very recently, an analogous supramolecular trimetallic species has been reported where the central Ir-based moiety is replaced by a Rh(III) complex [127]. This new system, $[(bpy)_2Ru(dpp)]_2RhCl_2{}^{5+}$ (**25**), was obtained coupling two Ru chromophoric units which play the role of P, through

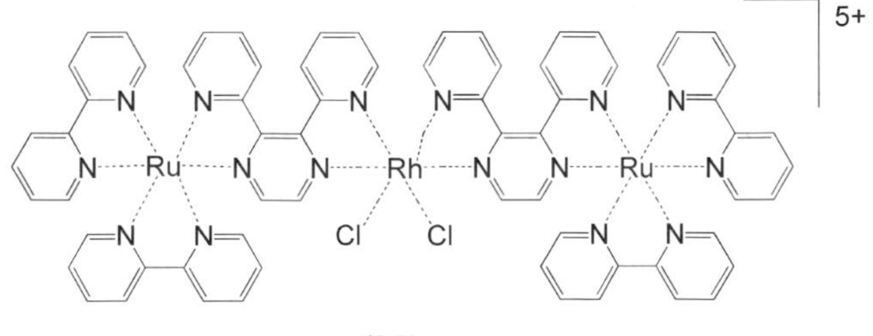

(25)

polyazine bridges (BL), to a central Rh core (C). In the presence of dimethylaniline (DMA) as sacrificial electron donor (D), **25** undergoes two-electron photoreduction at the rhodium center, producing the stable Rh(I) form $[(bpy)_2Ru(dpp)]_2Rh^{5+}$ via loss of two chlorides. This result is demonstrated by the observation that the spectroscopic changes associated with the photochemical reduction are identical with those seen in the electrochemical reduction experiments. On the basis of quenching experiments in the presence of different concentrations of DMA, the authors suggest that the monoreduced Rh(II) species can be formed by two alternative pathways following excitation of the peripheral Ru(II)-base chromophores: i) photoinduced electron transfer from the excited Ru(II)-based units followed by regeneration of the Ru(II)-based chromophores by oxidation of the sacrificial donor or ii) bimolecular quenching of the excited Ru(II)-based units by the sacrificial donor followed by reduction of the central Rh(III). No clear indication is given as to the mechanism for the formation of the stable two-electron-reduced Rh(I) product from the Rh(II) intermediate. According to the authors, the ability of $[(bpy)_2Ru(dpp)]_2RhCl_2{}^{5+}$ to undergo photoreduction at the rhodium center by multiple electrons and the fact that the photoreduced product $[(bpy)_2Ru(dpp)]_2Rh^{5+}$ is coordinatively unsaturated and thus available to interact with substrates are promising features in view of potential applications in light energy harvesting to produce fuels [127].

4 Rhodium Complexes as DNA Intercalators

4.1 Specific Binding to DNA and Photocleavage

The development and the study of transition metal complexes able to bind selectively to DNA sites, emulating the behavior of the DNA-binding proteins, ranks among the most fascinating and challenging issues in the field of current chemical and biological research [128–131]. This topic has been extensively investigated by Barton and coworkers, who devoted an impressive number of studies to the use of Rh(III) complexes with ligands containing nitrogen donors as DNA binding agents [129, 130]. Most of these studies center around complexes of the ligand 9,10-phenanthrenequinonediimine (phi) (**26**) [130, 132–138].

The phi Rh(III) complexes are indeed excellent DNA intercalators given the flat aromatic heterocyclic moiety of the phi ligand that deeply inserts and stacks in between the DNA base pairs (binding affinity constants range from 10^6–10^9 M^{-1}) [132, 139, 140]. The photophysical properties of phi Rh(III) complexes [50] have been discussed in Sect. 2.1. When bound to DNA, upon photoactivation with UV light, they are able to promote DNA strand cleav-

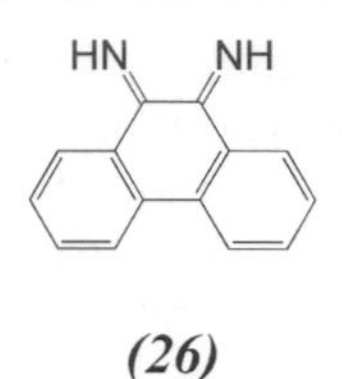

(26)

age, thanks to the outstanding oxidant properties of their excited states. The photocleavage ability offers a strategy for the use of these rhodium complexes as DNA targets. The approach used by the Barton's laboratory is the following: i) upon ultraviolet excitation, the excited state of DNA-bound rhodium complex promotes the scission of DNA sugar-phosphate backbone through oxidative degradation of the sugar moiety; ii) biochemical methods (e.g., gel electrophoresis) are used to determine where the strand scission occurred and therefore where, along the strand, the complex was bound. This method provides a powerful tool to mark specifically the sites of binding [129].

A variety of articles focused on DNA photocleavage by phi complexes containing different ligands in ancillary positions [130, 139–142]. The structural formula of the most extensively characterized complexes are reported below (**27, 28, 29, 30**).

Irradiation with UV light of $Rh(phen)_2(phi)^{3+}$ (**28**) and $Rh(bpy)(phi)_2{}^{3+}$ (**30**) intercalated in DNA leads to direct DNA strand scission with products

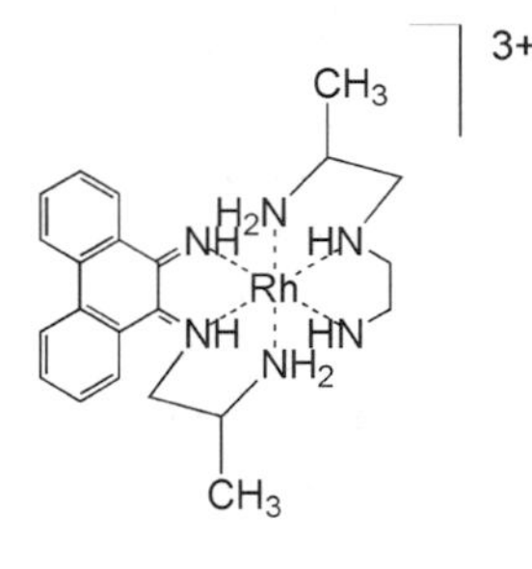

(27)

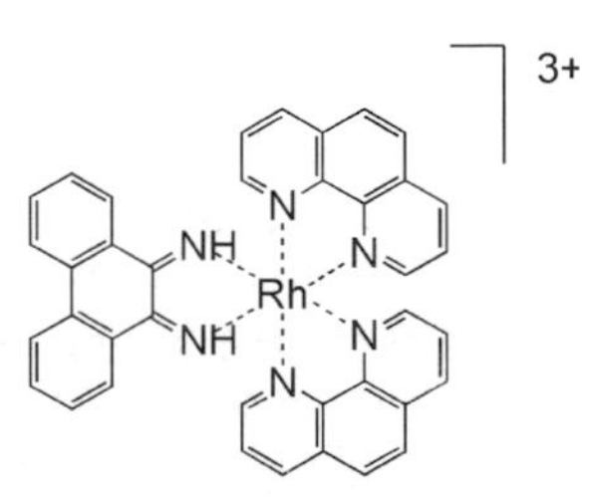

(28)

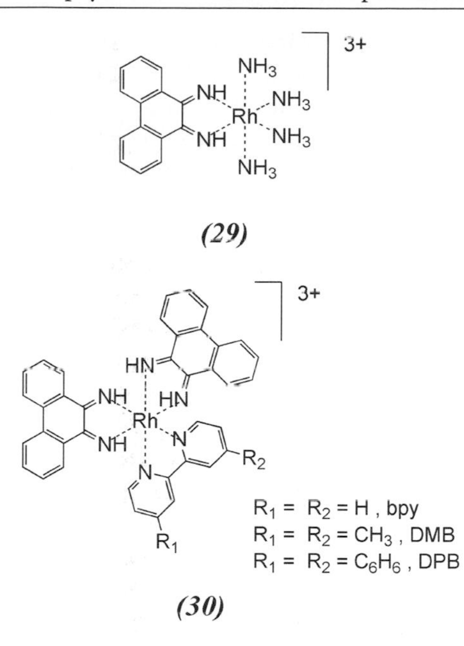

consistent with 3′-hydrogen abstraction from the deoxyribose sugar adjacent to the binding site [140, 141]. The photochemistry of these phi complexes intercalated in DNA has also been studied as a function of irradiation wavelength [143, 144]. Interestingly, the results showed that light of different wavelengths induces selectively different chemical reactions. In particular, the irradiation of the DNA-bound $Rh(phi)_2(L)^{3+}$ complexes with UV light ($\lambda = 313$ nm) leads, as discussed above, to direct scission of the DNA-sugar backbone [141, 144]. If instead the complexes are excited with low-energy light ($\lambda \geq 365$ nm) oxidative damage to the DNA bases is observed. The mechanism of these two photoprocesses is not discussed in great detail [130, 143]. It is proposed, however, which direct DNA scission takes place by hydrogen abstraction from the sugar by the phi ligand radical of a ligand-to-metal charge transfer (LMCT) state [130]. On the other hand, the oxidative damage is attributed to the population of a powerful oxidizing excited state (ILCT [50] or LC [143]) with longer wavelength light.

Photocleavage experiments have been used profitably for establishing how the phi complexes are associated to DNA [129, 130, 141]. Confirmation of site selectivity and greater structural definition were obtained later from high-resolution NMR studies [134–138, 145]. The important result is that all the phi complexes bind DNA noncovalently through intercalation in the major groove where the phi ligand is inserted between the base pairs so as to maximize stacking interactions. More recently a full crystal structure of Δ– Rh $((R,R)\text{-Me}_2\text{trien})_2(\text{phi})^{3+}$ ((R,R)-Me_2trien = $2R,9R$-diamino-4,7-diazadecane) bound to a DNA octamer provided a direct evidence of the intercalation through the major groove [146]. A series of systematic NMR and photocleavage studies clearly showed that the binding of complexes containing different ancillary ligands occurs at a different specific DNA sequence. This site specificity results from both shape-selective steric interactions as well as stabilizing van der Waals and hydrogen bonding contacts. In particular, $Rh(NH_3)_4phi^{3+}$ and related amine complexes bind to $d(TGGCCA)_2$ duplex through hydrogen bonding between the ancillary amine ligands and DNA bases [130, 136]. Evidence for specific intercalation was found also for $Rh(phen)_2phi^{3+}$ in the hexanucleotide $d(GTCGAC)_2$ [134]. In this case Barton proposed that the site specificity was based upon shape-selection. Since the phenanthroline ligands provide steric bulk above and below the plane of the phi ligand, the stacking occurs at sites which are more open in the major grove. The most striking example of site-specific recognition by shape selection with bulky ancillary ligands was found for $Rh(DPB)_2phi^{3+}$ (DPB = 4,4′-diphenylbpy) [140]. For all the complexes studied enantioselectivity favoring the intercalation of the Δ-isomer was observed [130, 147]. Further control of sequence specificity has been achieved by using derivatives of $Rh(phen)(phi)_2{}^{3+}$ complexes where the nonintercalating phenanthroline ligand has been functionalized with pendant guanidinium group or with short oligopeptides [148, 149]. For metal-peptide complexes photocleavage experiments showed that the polypeptide chain is essential in directing the complex to a specific DNA sequence [149].

Among the rhodium intercalators explored as probes of DNA structure, Barton selected the $Rh(bpy)_2chrysi^{3+}$ (chrysi = 5,6-chrysenequinone diimine, **31**) complex as an ideal candidate for mismatches recognition [150–152].

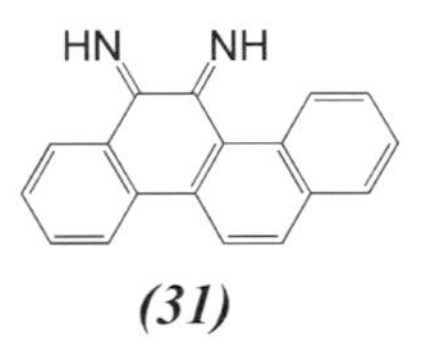

(31)

The specific recognition is based on the size of the intercalating ligand: the chrysene ring system is too large to intercalate in normal B-form DNA but it can do so at destabilized mismatch sites. The authors point out that sterically demanding intercalators such as $Rh(bpy)_2chrysi^{3+}$ may have application both in mutation detection systems and as mismatch-specific chemotherapeutic agents.

Recently mixed-metal trimetallic complexes have been designed and studied by Brewer to obtain supramolecular system capable of DNA photocleavage [153, 154]. These complexes of general formula $[\{(bpy)_2M(dpp)\}_2RhCl_2](PF_6)_5$ with M = Ru(II) or Os(II) couple ruthenium or osmium chromophoric units to a central rhodium(III) core. When excited with visible light into their intense MLCT bands, these complexes exhibit DNA photocleavage property. The authors discussed the role of the supramolecular architecture and in particular of the rhodium(III) unit on the photoreactivity

4.2 Rh(III) Complexes in DNA-Mediated Long-Range Electron Transfer

DNA-mediated electron transfer has been an active and much debated topic of research [155]. Several articles have dealt with DNA-mediated photoinduced electron transfer reactions involving metal complexes as photoactive units [156]. In this context, of particular interest are the studies of Barton and coworkers that used rhodium(III) intercalator complexes not only as ground-state but also as excited-state electron acceptor in electron transfer (ET) reactions through the DNA [144].

4.2.1 Rh(III) Complexes as Acceptors in Electron Transfer Reactions

In 1993, Murphy et al. [157]. reported the surprising result that an efficient and rapid photoinduced electron transfer occurs over a large separation distance (> 40 Å) between DNA metallointercalators that are covalently tethered to opposite 5′-ends of a 15-base pair DNA duplex (Fig. 13).

In this oligomeric assembly $Ru(phen)_2dppz^{2+}$ (dppz = dipyridophenazine) plays the role of excited electron donor and the $Rh(phi)_2(phen)^{3+}$ is the electron acceptor. Both donor and acceptor bind to DNA with high affinities ($> 10^6\ M^{-1}$) by intercalation through the dppz [158, 159] and phi ligands, re-

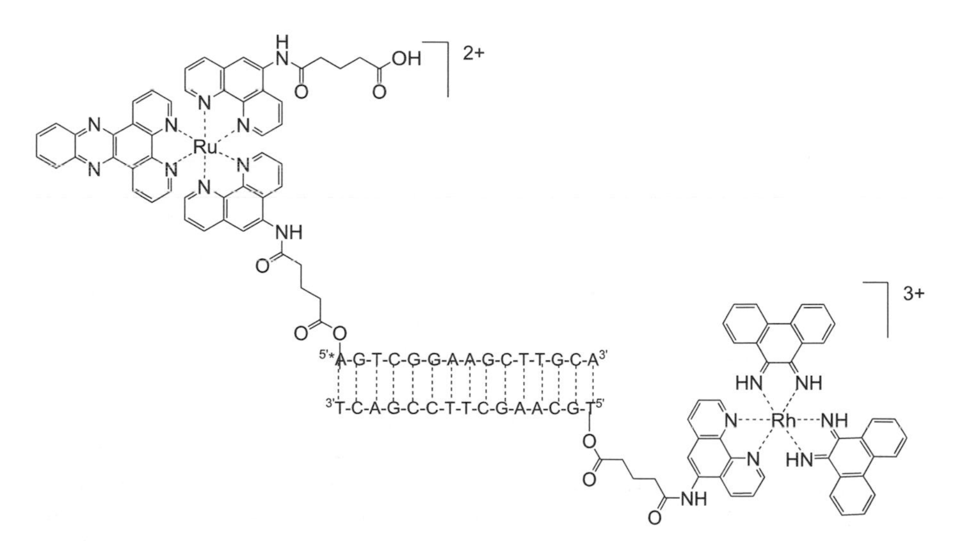

Fig. 13 A 15-base pair DNA duplex carrying covalently tethered Ru(II) and Rh(III) intercalators

spectively. A clear advantage of the tethered Ru/DNA/Rh system is that both the donor and acceptor are covalently held in a well-defined fixed distance range. On the other hand, the report of Murphy et al. was limited by the exclusive use of steady-state emission spectroscopy. A lower limit for the photoinduced electron transfer rate ($> 3 \times 10^9\ s^{-1}$) has been obtained measuring the quenching of the Ru(II) metal-to-ligand charge transfer (MLCT) emission by the tethered Rh(III) acceptor.

In a subsequent investigation, untethered Ru/DNA/Rh and related systems were studied by Barton et al. [160] using ultrafast laser spectroscopy. The study was focused mainly on the system shown in Fig. 14 constituted by Δ-$Ru(phen)_2dppz^{2+}$ as excited donor, Δ-$Rh(phi)_2bpy^{3+}$ as acceptor intercalated in the calf thymus DNA with the aim to determine the rate of excited-state electron transfer (k_{et}) that occurs from the lowest-lying MLCT state of the Ru donor, and the recombination electron transfer reaction (k_{rec}).

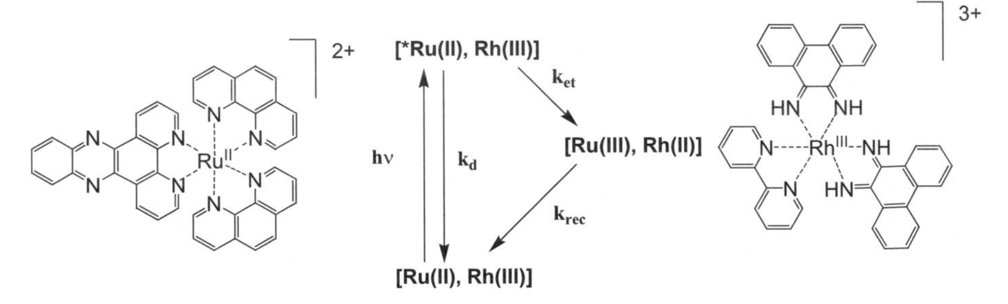

Fig. 14 Photoinduced electron transfer processes taking place between $Ru(phen)_2dppz^{2+}$ and $Rh(phi)_2bpy^{3+}$ DNA intercalators

Efficient and rapid quenching of luminescence of the Ru complex in the presence of Rh complex, even at surprisingly low acceptor loading on the DNA duplex was observed. All the experimental observations were consistent with complete intercalation of the donor and acceptor in DNA. A comparative experiment employing $Ru(NH_3)_6^{3+}$ complex as electron acceptor, clearly indicates that much less efficient quenching is observed when the quencher is groove bound rather intercalated. To deepen the understanding of the mechanism of the electron transfer processes, the authors examined the photoinduced charge separation (k_{et}) and recombination electron transfer (k_{rec}) reactions on the picosecond time scale by monitoring both the kinetics of the emission decay and the kinetics of the recovery of ground state absorption of Ru(II) donor (Fig. 14). Time-correlated single photon counting failed to detect the lifetime of the excited state, clearly indicating that luminescent quenching by electron transfer proceeds faster ($k_{et} > 3 \times 10^{10}\ s^{-1}$) than the time resolution of the instrument (ca. 50 ps). Ultrafast transient absorption measurements, on the other hand, revealed bleaching of the MLCT band of the Ru(II) complex in a picosecond time scale, assigned by the authors to the

Table 1 Rate constants for charge recombination following electron transfer from DNA-bound, photoexcited donors to Δ-Rh(phi)$_2$(bpy)$^{3+}$ [a]

Donor	DNA	k_{rec} [10^9 s^{-1}]	$-\Delta G°$ [V]
Δ-Ru(phen)$_2$(dppz)$^{2+}$	Calf thymus	9.2	1.66
rac-Ru(bpy)$_2$(dppz)$^{2+}$	Calf thymus	7.1	1.69
Δ-Ru(dmp)$_2$(dppz)$^{2+}$	Calf thymus	11	1.59
Δ-Ru(phen)$_2$(F$_2$dppz)$^{2+}$	Calf thymus	7.7	1.68
rac-Ru(phen)$_2$(Me$_2$dppz)$^{2+}$	Calf thymus	9.2	1.67
Δ-Os(phen)$_2$(dppz)$^{2+}$	Calf thymus	11	1.21
Λ-Ru(phen)$_2$(dppz)$^{2+}$	Calf thymus	4.5	
Δ-Ru(phen)$_2$(dppz)$^{2+}$	Poly-d(AT)	7.4	
Δ-Ru(phen)$_2$(dppz)$^{2+}$	Poly-d(GC)	0.21	

[a] Based on [144]

presence of Ru(III) oxidized donor. The rate constants for charge recombination process (k_{rec}) were obtained from the decay of this signal. The data for seven donor–acceptor pairs are given in Table 1. Within this series, the driving force ($\Delta G°$) is comparable but the donors vary with respect to intercalating ligand, ancillary ligands, chirality, and metal center. Despite such a range of chemical properties, the rate observed is centered around 10^{10} s^{-1}.

A significant difference in rate is observed, however, when the absolute configuration of the donor is varied. For the right-handed Δ-Ru(phen)$_2$dppz^{2+} the value is 2.5 times higher with respect the left-handed enantiomer indicating a deeper stacking of this complex into the double helix. This result, according to the authors, clearly suggests that the electron transfer process required the intervening aromatic base pairs. The notion of highly efficient ET through the stack of DNA base is also strongly supported by the finding that the largest change in electron transfer rate is observed when the sequence of the DNA bridge is changed: for the same donor and acceptor reactants the rate with poly d(AT) is 30 times higher than with poly d(GC). This is an important result that indicates that the π-stacked bases of the DNA provide an effective pathway for electron transfer reactions. However, the crucial point of this study that involves an untethered Ru/DNA/Rh system concerns the distances over which fast ET occurs. The question is: do the donor and acceptor complexes contact each other, or does electron transfer occur at long range? Two models were considered by the authors to interpret the experimental results: i) a cooperative binding model with ET over short D–A distance and ii) a random binding model with ET over long distances. On the basis of DNA photocleavage experiments, the first hypothesis was reject in favor of a rapid long range ET with a shallow distance dependence [160]. On the other hand, soon thereafter, Barbara [161] reinterpreted the experimental results on a quantitative basis using computational simulation procedures and demonstrated the failure of long-distance electron transfer model to account for the data. Concurrently, Tuite and coworkers [162] arrived at a similar conclusion for the ET quenching of $Ru(phen)_2dppz^{2+}$ emission in a very similar untethered DNA/metallointercalator system. In summary, considerable controversy persists in the estimates of the distances over which fast ET may occur in such type of untethered systems [144].

4.2.2 Long Range Oxidative DNA Damage by Excited Rh(III) Complexes

It is well known from a large variety of experimental studies and calculations that guanine (G) is the most easily oxidized of the nucleic acid bases [144, 163, 164]. Barton and coworkers have extensively exploited the ability of Rh(III)-phi complexes to induce oxidative damage specifically at the 5′-G of the 5′-GG-3′ doublets, when irradiated with low-energy light. A first investigation was carried out using a 15-base duplex (Rh-DNA) which possesses an end-tethered $Rh(phi)_2bpy^{3+}$ complex in one strand and two 5′-GG-3′ sites in the complementary strand. (Fig. 15). The peculiarity of this Rh-DNA assembly is that the rhodium complex is spatially separated in a well-defined manner from the potential sites of oxidation. Damage to DNA was demonstrated to occurred as a result of excitation of the intercalated rhodium complex, followed by long-range (30–40 Å) electron transfer through the DNA base pair stack [165]. The strategy used to analyze the mechanism is illustrated in Fig. 16.

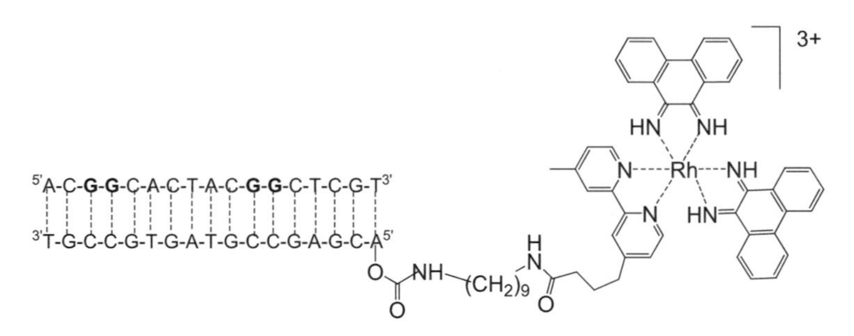

Fig. 15 A15-base duplex with an end-tethered $Rh(phi)_2bpy^{3+}$ complex in one strand and two 5′-GG-3′ sites in the complementary strand [165]

The Rh-DNA assembly was first irradiated at 313 nm to induce direct strand cleavage. This photocleavage step marks the site of intercalation, and permits determination of the distance separating the rhodium complex from potential sites of damage. Rh-DNA samples were then irradiated with low energy light at 365 nm, treated with hot piperidine, which promotes strand cleavage at the damaged sites, and examined by gel electrophoresis. This treatment reveals the position and yield of damage. The results clearly in-

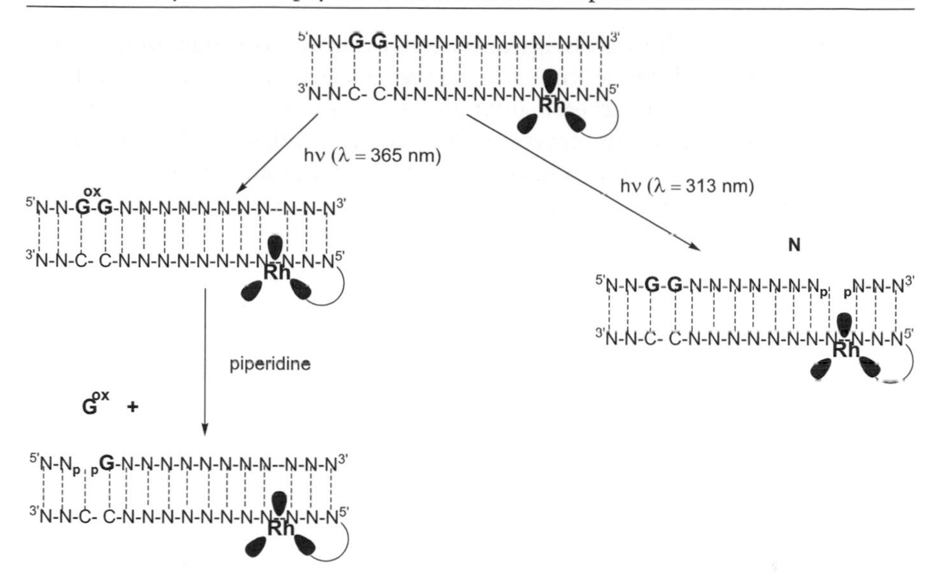

Fig. 16 Use of the tethered Rh(III) complex to (i) mark the site of intercalation by direct strand cleavage (313-nm irradiation) and (ii) promote damage via long-range electron transfer (365-nm irradiation) [165]

dicated that both the proximal and distal 5′-GG-3′ doublets were equally damaged and the reaction was intraduplex. Two possible mechanisms for this process were discussed: i) concerted long-range electron transfer; and ii) oxidation of a base near the intercalated Rh acceptor followed by hole migration to the two GG sites. Sensitivity of the reaction to the intervening base pair stack was also observed. In subsequent studies, oxidation has been reported at sites that are up to 200 Å away from the site of intercalation of the photoactive rhodium complex [166].

The photooxidant properties of the phi rhodium (III) complexes have also been used to repair thymine dimers [167, 168], the most common photo-

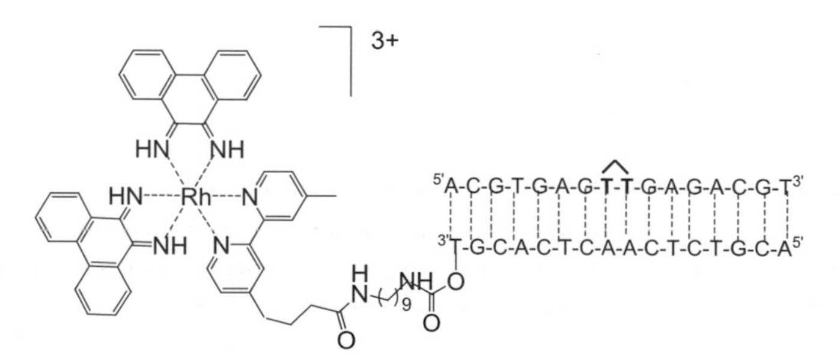

Fig. 17 DNA duplex containing a thymine dimer with tethered Rh(III) complex for photoinduced repair studies [167, 168]

chemical lesion in DNA. Investigations of photoinitiated repair of duplexes containing a single thymine dimer lesion were carried out with visible light (400 nm) using both nontethered and tethered complexes (Fig. 17).

The quantum yield for photorepair with a Rh(III)-tethered complex is substantially (about ca. 30 fold) reduced compared to the noncovalently bound complex. Since the repair efficiency does not appear to be very sensitive to the distance between intercalated rhodium complex and the thymine dimer, the authors suggest that the observed disparity likely results from differences in π-stacking. In addition, evidences that the repair efficiency diminished with disruption of the intervening π-stack confirm that the DNA helix mediates this long-range oxidative repair reaction.

5 Conclusion

A large number of rhodium(III) polypyridine complexes and their cyclometalated analogues have been investigated from the viewpoint of photochemistry, photophysics and of their possible applications.

As mononuclear species, Rh(III) polypyridine complexes display interesting photophysical properties, with lowest excited states of LC type for tris bis-chelated species, and increasing role of MC states for mixed-ligand halo-polypyridine species. In Rh(III) cyclometalated complexes, the covalent character of the C – Rh bonds makes the excited state classification less clearcut, with strong mixing of LC, MLCT, and LLCT character.

Many polynuclear and supramolecular systems containing Rh(III) polypyridine and related units have been synthesized and studied, taking advantage of the favorable properties of these units as good electron acceptors and strong photo-oxidants. In particular, Ru(II)-Rh(IIII) dyads have been actively investigated for the study of photoinduced electron transfer, with specific interest in driving force, distance, and bridging ligand effects. A limited number of supramolecular systems of higher nuclearity have also been produced. Among these, of particular interest are trinuclear species containing rhodium dihalo polypyridine units, which can act as two-electron storage components thanks to their Rh(III)/Rh(I) redox behavior.

Finally, a large amount of work has been devoted to the use of Rh(III) polypyridine complexes as intercalators for DNA. In this role, they have shown a very versatile behavior, being used for direct strand photocleavage marking the site of intercalation, to induce long-distance photochemical damage or dimer repair, or to act as electron acceptors in long-range electron transfer processes.

Acknowledgements Financial support from MUR (PRIN 2006) is gratefully acknowledged.

References

1. Carassiti V, Balzani V (1970) Photochemistry of Coordination Compounds. Academic Press, New York
2. Kalyanasundaram K (1992) Photochemistry of Polypyridine and Porphyrin Complexes. Academic Press, London
3. Balzani V, Juris A, Venturi M, Campagna S, Serroni S (1996) Chem Rev 96:759
4. Hannon MJ (1997) Coord Chem Rev 162:477
5. Scandola F, Chiorboli C, Indelli MT, Rampi MA (2001) Covalently linked systems containing metal complexes. In: Balzani V (ed) Electron Transfer in Chemistry, Vol III, Chap 2.3. Wiley-VCH, Weinheim, p 337
6. Chiorboli C, Indelli MT, Scandola F (2005) Top Curr Chem 257:63
7. Ford PC (1983) J Chem Ed 60:829
8. Ford PC, Wink D, DiBenedetto J (1983) Prog Inorg Chem 40:213
9. Miller DB, Miller PK, Kane-Maguire NAP (1983) Inorg Chem 22:3831
10. Frink ME, Ford PC, Skibsted LH (1984) Acta Chem Scand A38:795
11. Skibsted LH, Hancock MP, Magde D, Sexton DA (1984) Inorg Chem 23:3735
12. Skisbsted LH (1985) Coord Chem Rev 64:343
13. Kane-Maguire NAP, Wallace KC, Cobranchi DP, Derrick JM, Speece DG (1986) Inorg Chem 25:2101
14. Skibsted LH, Hancock MP, Magde D, Sexton DA (1987) Inorg Chem 26:1708
15. McClure LJ, Ford PC (1992) J Phys Chem 96:6640
16. Carlos RM, Frink ME, Tfouni E, Ford PC (1992) Inorg Chim Acta 193:159
17. Islam K, Ikeda AN, Nozaki KT, Ohno T (1998) J Chem Phys 109:4900
18. Forster LS, Rund JV (2003) Inorg Chem Commun 6:78
19. Coppens P, Gerlits O, Vorontsov II, Kovalevsky AY, Chen Y-S, Graber T, Novozhilova IV (2004) Chem Commun, p 2144
20. Miskowski VM, Rice SF, Gray HB, Milder SJ (1993) J Phys Chem 97:4277
21. Bradley PM, Bursten BE, Turro C (2001) Inorg Chem 40:1376
22. Cotton FA, Lin C, Murillo CA (2001) Acc Chem Res 34:759
23. Lo Schiavo S, Serroni S, Puntoriero F, Tresoldi G, Piraino P (2002) Eur J Inorg Chem 79
24. Cooke MW, Hanan GS, Loiseau F, Campagna S, Watanabe M, Tanaka Y (2005) Angew Chem Int Ed 44:4881
25. Heyduk AF, Macintosh AM, Nocera DG (1999) J Am Chem Soc 121:5023
26. Heyduk AF, Nocera DG (2001) Science 293:1639
27. Gray TG, Nocera DG (2005) Chem Commun, p 1540
28. Hillis JE, De Armond MK (1971) J Lumin 4:273
29. Carstens DHW, Crosby GA (1970) J Mol Spectrosc 34:113
30. Crosby GA, Elfring WH Jr (1976) J Phys Chem 80:2206
31. De Armond MK, Hillis JE (1971) J Chem Phys 54:2247
32. Humbs W, Yersin H (1996) Inorg Chem 35:2220
33. Humbs W, Strasser J, Yersin H (1997) J Lumin 72–74:677
34. Indelli MT, Carioli A, Scandola F (1984) J Phys Chem 88:2685
35. Bolletta F, Rossi A, Barigelletti F, Dellonte S, Balzani V (1981) Gazz Chim Ital 111:155
36. Ohno T, Kato S (1984) Bull Chem Soc Jpn 57:3391
37. Nishizawa M, Suzuki TM, Sprouse S, Watts RJ, Ford PC (1984) Inorg Chem 23:1837
38. Indelli MT, Scandola F (1990) Inorg Chem 29:3056
39. Brozik JA, Crosby GA (2005) Coord Chem Rev 249:1310
40. Broomhead JA, Grumley W (1968) Chem Commun, p 1211

41. Muir MM, Huang WL (1973) Inorg Chem 12:1831
42. Loganathan D, Rodriguez JH, Morrison H (2003) J Am Chem Soc 125:5640
43. Menon EL, Perera R, Navarro M, Kuhn RJ, Morrison H (2004) Inorg Chem 43:5373–5381
44. Huang WL, Lee JR, Shi SY, Tsai CY (2003) Trans Met Chem 28:381
45. Tseng MC, Li FK, Huang JH, Su WL, Wang SP, Huang WL (2006) Inorg Chim Acta 359:401
46. Calvert JM, Caspar JV, Binstead RA, Westmoreland TD, Meyer TJ (1982) J Am Chem Soc 104:6620
47. Frink ME, Sprouse SD, Goodwin HA, Watts RJ, Ford PC (1988) Inorg Chem 27:1283
48. Watts RJ, Van Houten J (1978) J Am Chem Soc 100:1718
49. Westra J, Glasbeek M (1992) J Lumin 53:92
50. Turro C, Evenzahav A, Bossmann SH, Barton JK, Turro NJ (1996) Inorg Chim Acta 243:101
51. Burke HM, Gallagher JF, Indelli MT, Vos JG (2004) Inorg Chim Acta 357:2989
52. Shinozaki K, Takahashi N (1996) Inorg Chem 35:3917
53. Fujita E, Brunschwig BS, Creutz C, Muckerman JT, Sutin N, Szalda D, van Eldik R (2006) Inorg Chem 45:1595
54. Martin B, McWhinnie WR, Waind GM (1961) J Inorg Nucl Chem 23:207
55. Chou M, Creutz C, Mahajan D, Sutin N, Zipp AP (1982) Inorg Chem 21:3989
56. Mulazzani QG, Venturi M, Hoffman MZ (1982) J Phys Chem 86:242
57. Schwarz HA, Creutz C (1983) Inorg Chem 22:707
58. Yan SG, Brunschwig BS, Creutz C, Fujita E, Sutin N (1998) J Am Chem Soc 120:10553
59. Maestri M, Balzani V, Deuschel-Cornioley C, von Zelewsky A (1992) Adv Photochem 17:1
60. Colombo MG, Brunold TC, Riedener T, Guedel HU, Fortsch M, Buergi H-B (1994) Inorg Chem 33:545
61. Constable EC, Leese TA (1990) Polyhedron 9:1613
62. Maeder U, von Zelewsky A, Stoeckli-Evans H (1992) Helv Chim Acta 75:1320
63. Maestri M, Sandrini D, Balzani V, Mader U, von Zelewsky A (1987) Inorg Chem 26:1323
64. Ohsawa Y, Sprouse S, King KA, De Armond MK, Hanck KW, Watts RJ (1987) J Phys Chem 91:1047
65. Zilian A, Maeder U, von Zelewski A, Guedel HU (1989) J Am Chem Soc 111:3855
66. Zilian A, Guedel HU (1991) Coord Chem Rev 111:33
67. Ghizdavu L, Lentzen O, Schumm S, Brodkorb A, Moucheron C, Kirsch-De Mesmaeker A (2003) Inorg Chem 42:1935
68. Hay PJ (2002) J Phys Chem A 106:1634
69. Polson M, Ravaglia M, Fracasso S, Garavelli M, Scandola F (2005) Inorg Chem 44:1282
70. Barigelletti F, Sandrini D, Maestri M, Balzani V, von Zelewsky A, Chassot L, Jolliet P, Maeder U (1988) Inorg Chem 27:3644
71. Lo KK-W, Li C-K, Lau K-W, Zhu N (2003) Dalton Trans 24:4682
72. Sandrini D, Maestri M, Balzani V, Mader U, von Zelewsky A (1988) Inorg Chem 27:2640
73. Calogero G, Giuffrida G, Serroni S, Ricevuto V, Campagna S (1995) Inorg Chem 34:541
74. Didier P, Ortmans I, Kirsch-De Mesmaeker A, Watts RJ (1993) Inorg Chem 32:5239
75. Van Diemen JH, Hage R, Haasnoot JG, Lempers HEB, Reedijk J, Vos JG, De Cola L, Barigelletti F, Balzani V (1992) Inorg Chem 31:3518

76. Ortmans I, Didier P, Kirsch-De Mesmaeker A (1995) Inorg Chem 34:3695
77. Petitjean A, Puntoriero F, Campagna S, Juris A, Lehn J-M, (2006) Eur J Inorg Chem 3878
78. Kew G, Hanck K, DeArmond MK (1975) J Phys Chem 79:1828
79. Ballardini R, Varani G, Balzani V (1980) J Am Chem Soc 102:1719
80. Indelli MT, Ballardini R, Scandola F (1984) J Phys Chem 88:2547
81. Lehn J-M, Sauvage J-P (1977) Nouv J Chim 1:449
82. Chan S-F, Chou M, Creutz C, Matsubara T, Sutin N (1981) J Am Chem Soc 103:369
83. Creutz C, Keller AD, Sutin N, Zipp A P (1982) J Am Chem Soc 104:3618
84. Indelli MT, Bignozzi CA, Harriman A, Schoonover JR, Scandola F (1994) J Am Chem Soc 116:3768
85. Yoshimura A, Nozaki K, Ohno T (1997) Coord Chem Rev 159:375
86. Furue M, Hirata M, Kinoshita S, Kushida T, Kamachi M (1990) Chem Lett 2065
87. Collin J-P, Gavina P, Heitz V, Sauvage J-P (1998) Eur J Inorg Chem :1
88. Welter S, Salluce N, Belser P, Groeneveld M, De Cola L (2005) Coord Chem Rev 249:1360
89. Welter S, Lafolet F, Cecchetto E, Vergeer F, De Cola L (2005) Chem Phys Chem 6:2417
90. Barigelletti F, Flamigni I (2000) Chem Soc Rev 29:1
91. Indelli MT, Scandola F, Flamigni L, Collin J-P, Sauvage J-P, Sour A (1997) Inorg Chem 36:4247
92. Indelli MT, Chiorboli C, Flamigni L, De Cola L, Scandola F (2007) Inorg Chem (in press)
93. McConnell HM (1961) J Chem Phys 35:508
94. Newton MD (1991) Chem Rev 91:767
95. Paddon-Row MN (2001) Covalently linked systems based on organic components. In: Balzani V (ed) Electron Transfer in Chemistry, Vol III, Chap 2.1. Wiley-VCH, Weinheim, p 179
96. Helms A, Heiler D, McLendon G (1992) J Am Chem Soc 114:6227
97. Weiss EA, Ahrens MJ, Sinks LE, Gusev AV, Ratner MA, Wasielewsi MR (2004) J Am Chem Soc 126:5577
98. Wold DJ, Haag R, Rampi MA, Frisbie CD (2002) J Phys Chem B 106:2814
99. Toutounji MM, Ratner MA (2000) J Phys Chem A 104:8566
100. Tour JM, Lamba JJS (1993) J Am Chem Soc 115:4935
101. Tsuzuki K, Tanabe J (1991) J Phys Chem 95:139
102. Kalyanasundaram K, Graetzel M, Nazeeruddin MK (1992) J Phys Chem 96:5865
103. Lee J-D, Vrana LM, Bullock ER, Brewer KJ (1998) Inorg Chem 37:3575
104. Kew G, DeArmond K, Hanck K (1974) J Phys Chem 78:727
105. Collin J-P, Harriman A, Heitz V, Odobel F, Sauvage J-P (1994) J Am Chem Soc 116:5679
106. Wasielewski MR (1992) Chem Rev 92:435
107. Gust D, Moore TA, Moore AL (2001) Covalently linked systems containing porphyrin units. In: Balzani V (ed) Electron Transfer in Chemistry, Vol III, Chap 2.2. Wiley-VCH, Weinheim, p 273
108. Larson SL, Elliott CM, Kelley DF (1995) J Phys Chem 99:6530
109. Huynh MHV, Dattelbaum DM, Meyer TJ (2005) Coord Chem Rev 249:457
110. Baranoff E, Collin J-P, Flamigni L, Sauvage J-P (2004) Chem Soc Rev 33:147
111. Ronco SE, Thompson DW, Gahan SL, Petersen JD (1998) Inorg Chem 37:2020
112. Indelli MT, Polo E, Bignozzi CA, Scandola F (1991) J Phys Chem 95:3889
113. Kleverlaan CJ, Indelli MT, Bignozzi CA, Pavanin LA, Scandola F, Hasselman GM, Meyer GJ (2000) J Am Chem Soc 122:2840

114. Balzani V, Scandola F (1991) Supramolecular Photochemistry. Horwood, Chichester
115. Konduri R, Ye H, MacDonnell FM, Serroni S, Campagna S, Rajeshwar K (2002) Angew Chem Int Ed 41:3185
116. Chiorboli C, Fracasso S, Ravaglia M, Scandola F, Campagna S, Wouters K, Konduri R, MacDonnell F (2005) Inorg Chem 44:8368
117. Heyduk AF, Nocera DG (2001) Science 293:1639
118. Esswein A, Veige A, Nocera D (2005) J Am Chem Soc 127:16641
119. Chang CC, Pfennig B, Bocarsly AB (2000) Coord Chem Rev 208:33
120. Watson DF, Tan HS, Schreiber E, Mordas CJ, Bocarsly AB (2004) J Phys Chem A 108:3261
121. Rau S, Schfer B, Gleich D, Anders E, Rudolph M, Friedrich M, Gorls H, Henry W, Vos JG (2006) Angew Chem Int Ed 45:6215
122. Ozawa H, Haga M, Sakai K (2006) J Am Chem Soc 128:4926
123. Molnar SM, Nallas G, Bridgewater JS, Brewer KJ (1994) J Am Chem Soc 116:5206
124. Nallas GNA, Jones SW, Brewer KJ (1996) Inorg Chem 35:6974
125. Molnar SM, Jensen GE, Vogler LM, Jones SW, Laverman L, Bridgewater JS, Richter MM, Brewer KJ (1994) J Photochem Photobiol A Chem 80:315
126. Swavey S, Brewer KJ (2002) Inorg Chem 41:4044
127. Elvington M, Brewer KJ (2006) Inorg Chem 45:5242
128. Barton JK (1986) Science 233:727
129. Pyle AM, Barton JK (1990) In: Lippard SJ (ed) Progress in Inorganic Chemistry: Bioinorganic Chemistry, Vol 38. Wiley, New York, p 413
130. Erkkila KE, Odom DT, Barton JK (1999) Chem Rev 99:2777
131. Jennette KW, Lippard SJ, Vassiliades GA, Bauer WR (1974) Proc Natl Acad Sci USA 71:3839
132. Crotz AH, Hudson BP, Barton JK (1993) J Am Chem Soc 115:12577
133. Pyle AM, Chiang MY, Barton JK (1990) Inorg Chem 29:4487
134. David SS, Barton JK (1993) J Am Chem Soc 115:2984
135. Crotz AH, Kuo LY, Barton JK (1993) Inorg Chem 32:5963
136. Collins J-C, Shield JK, Barton JK (1994) J Am Chem Soc 116:9840
137. Crotz AH, Barton JK (1994) Inorg Chem 33:1940
138. Hudson BP, Barton JK (1998) J Am Chem Soc 120:6877
139. Crotz AH, Kuo LY, Shields TP, Barton JK (1993) J Am Chem Soc 115:3877
140. Sitlani A, Dupureur C, Barton JK (1993) J Am Chem Soc 115:12589
141. Sitlani A, Long EC, Pyle AM, Barton JK (1992) J Am Chem Soc 114:2303
142. Kisko J, Barton JK (2000) Inorg Chem 39:4942
143. Turro C, Hall DB, Chen W, Zuilhof H, Barton JK, Turro NJ (1998) J Phys Chem A 102:5708
144. Holmlin RE, Dandliker PJ, Barton KJ (1997) Angew Chem Int Ed 36:2714
145. Hudson BP, Dupureur CM, Barton JK (1995) J Am Chem Soc 117:9379
146. Kielkopf CL, Erkkila KE, Hudson BP, Barton JK, Rees DC (2000) Nat Struct Biol 7:117
147. Pyle AM, Morii T, Barton JK (1990) J Am Chem Soc 112:9432
148. Sardesai NJ, Zimmermann, Barton JK (1994) J Am Chem Soc 116:7502
149. Terbrueggen RH, Johann T W, Barton JK (1998) Inorg Chem 37:6874
150. Jackson BA, Barton JK (1997) J Am Chem Soc 119:12986
151. Jackson BA, Alekseyev VY, Barton JK (1999) Biochemistry 38:4655
152. Kisko JL, Barton JK (2000) Inorg Chem 39:4942
153. Holder AA, Swavey A, Brewer KJ (2004) Inorg Chem 43:303
154. Swavey S, Brewer KJ (2002) Inorg Chem 41:6196

155. Lewis FD (2001) In: Balzani V (ed) Electron Transfer in Chemistry, Vol III, Chap 1.5. Wiley/VCH, Weinheim, p 105
156. Stemp EDA, Barton JK (1996) In: Sigel A, Sigel H (eds) Metal Ions in Biological systems, Vol 33. Marcel Dekker, New York, p 325
157. Murphy CJ, Arkin MR, Jenkins Y, Ghatlia ND, Bossmann S, Turro NJ, Barton JK (1993) Science 262:1025
158. Friedman AE, Chambron JC, Sauvage JP, Turro NJ, Barton JK (1990) J Am Chem Soc 112:4960
159. Hort C, Lincoln P, Norden B (1993) J Am Chem Soc 115:3448
160. Arkin MR, Stemp EDA, Holmlin RE, Barton JK, Hormann A, Olson EJC, Barbara PF (1996) Science 273:475
161. Olson EJC, Hu D, Hormann A, Barbara PF (1997) J Phys Chem B 101:299
162. Lincoln P, Tuite E, Norden B (1997) J Am Chem Soc 119:1454
163. Saito I, Takayama M, Sugiyama H, Nakatani K (1995) J Am Chem Soc 117:6406
164. Kittler L (1980) J Electroanalyt Chem 116:503
165. Hall DB, Holmlin RE, Barton JK (1996) Nature 382:731
166. Nunez ME, Hall DB, Barton JK (1999) Chem Biol 6:85
167. Dandliker PJ, Holmlin RE, Barton JK (1997) Science 275:1465
168. Dandliker PJ, Nunez ME, Barton JK (1998) Biochemistry 37:6491

Author Index Volumes 251–280

Author Index Vols. 26–50 see Vol. 50
Author Index Vols. 51–100 see Vol. 100
Author Index Vols. 101–150 see Vol. 150
Author Index Vols. 151–200 see Vol. 200
Author Index Vols. 201–250 see Vol. 250

The volume numbers are printed in italics

Subject Index

Printing: Krips bv, Meppel
Binding: Stürtz, Würzburg